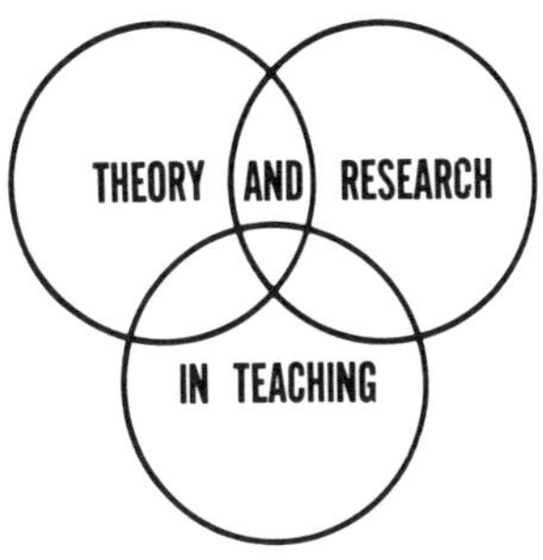

Teachers College, Columbia University
Arno A. Bellack, editor

Recent years have witnessed a resurgence of interest on the part of educational researchers in the teaching process. Volumes in the *Theory and Research in Teaching* series report significant studies of instructional procedures in a variety of educational settings, at various organizational levels in the schools, and in many of the subjects included in the curriculum. These studies present fresh perspectives on teaching both to educational researchers and to practitioners in the schools.

Patterns of Verbal Communication in Mathematics Classes

JAMES TAYLOR FEY

TEACHERS COLLEGE PRESS

Teachers College, Columbia University
New York, New York

Library of Congress Catalog Card Number: 74-103135
SBN No. 8077-1342

Manufactured in the United States of America

ACKNOWLEDGMENTS

The author would like to acknowledge the valuable guidance and criticism of Arno Bellack, Paul Rosenbloom, and Myron Rosskopf, as well as the cooperation of teachers and staff of the Secondary School Mathematics Curriculum Improvement Study and its director, Howard Fehr. Particular thanks go to my wife, Nancy, who served as a sounding board for many ideas and helped with the tedious data analysis and editing the final report.

J. T. F.

PREFACE

The revolution in school mathematics is now a decade old. The original interest in curriculum innovation has been extended to an equally active investigation of the way children learn mathematics. This research emphasis rests on the assumption that once a teacher has determined the important mathematical ideas and how children learn them, effective teaching is a routine corollary. There is, however, a growing awareness that learning and teaching are not synonymous. Mathematical content proposals and learning theory must be translated into prescriptions for curriculum organization and teaching theory; for this, it is necessary to build mathematics teaching theory.

The basic assumption of this study is that constructing and implementing theory on instruction in mathematics will be aided in important ways by deeper knowledge of the way mathematics teaching is now done. The characteristic classroom behaviors of mathematics teachers and students—when abstracted from systematic classroom observations—will provide a basic set of concepts to be organized and interrelated by theory. Then, when theory on instruction is formulated, teacher education will be able to point toward specific behavioral modifications needed if teachers are to attain or approach the theoretical ideal. Unfortunately, empirical knowledge about teaching that would be useful in theory construction is presently limited.

This lack of knowledge about classroom interaction can hardly be attributed to lack of interest in the problem. It is due to at least two fundamental failings of most relevant research. First, studies of teaching that were conducted in the past confined themselves primarily to correlating teaching success with a wide variety of teacher variables, such as personality, education, and experience. Jacob Getzels and Philip Jackson have pointed out that the leap from these correlations to predictions of teaching performance omits two important parts of inquiry: precise description of the teaching process, and the attempt to understand the dynamics of that process.

Inquiry that moves from observation to description to attempts to understand, and only then to prediction, might be more fruitful than the present trial-and-error search for correlations between conceptually barren "predictor variables" and teachership.

Second, empirical methods for recording and analyzing classroom activities have developed slowly—primarily for technological reasons. As Robert Travers has observed:

> Crude types of theory have long been part of the literature of education, but the propositions they have involved have typically been intuitively derived and have not been based on carefully collected data.

We are able to measure many characteristics of teachers and to relate these measures to various outcomes of teaching. But the teacher traits most often studied are measured either before or after the instructional period: How teachers behave in the classroom as they try to communicate with and influence their students, and how students respond to different teaching styles. These dynamics of teacher–student interaction have only recently become the topic of careful research.

Classroom interaction is a complex process. There are many dimensions of activity which must be examined and conceptualized before being measured—too many for one study to manage effectively. The present study concentrated on the *verbal* activity of teachers and students in mathematics classes. Since successful mathematics instruction depends in large measure on effective verbal communication, results in this important dimension should contribute substantially to the construction of a theory of instruction in mathematics. The general objectives of this research were:

1. To develop reliable empirical methods for describing the form, function, content, and sequence of verbal utterances in mathematics classrooms.
2. Assuming such methods can be developed, to describe the patterns of verbal interaction which can be identified in typical mathematics classes.

The rationale and specific procedures of this study are outlined in Chapter 1. The system developed for analyzing classroom verbal activity is presented in Chapter II. Patterns of verbal interaction in selected classes are discussed in Chapter III. Conclusions and suggestions for future research constitute Chapter IV.

CONTENTS

1 **Purpose and Design of the Research** **1**
Sources of Theory on Teaching, 1
Significant Related Research, 5
Research Design, 10

2 **The System For Analyzing Behavioral Records** **14**
The Classroom as a Communication System, 14
The Unit of Analysis, 16
Coding Moves, 17
Reliability Test, 35

3 **Verbal Communication in Selected Mathematics Classes** **37**
Overview of Verbal Activity, 37
Sources of Verbal Activity, 38
Pedagogical Purposes of Verbal Activity, 43
Content and Mathematical Purpose of Verbal Activity, 51
Content and Mathematical Activity Cycles, 55
Subordinate Content and Mathematical Activity, 63
Logical Processes in Verbal Activity, 66
Summary, 70

4 **Summary and Recommendations** **71**
Rules of the Language Game in Mathematics Classes, 72
Further Development of the System of Analysis, 74
Research Suggested by this Study, 75
Other Perspectives on Classroom Verbal Activity, 76

Bibliography **78**

Appendix **82**

Footnotes **89**

TABLES

I Average Number of Moves and Lines, Moves Per Minute, Lines Per Minute, and Lines Per Move for Each Class and for All Classes Combined. 38
II Average Number of Lines Per Move, Moves Per Minute, and Lines Per Minute for Each Session of Each Class and for All Classes Combined. 39
III Per cent of Lines and Moves for Teachers and Students in Each Class and in All Classes Combined. 40
IV Per cent of Lines and Moves for Teachers and Students in Each Class, Each Session. 41
V Comparison of Rates of Verbal Activity in Mathematics and Social Studies Classes. 42
VI Per cent of Moves and Lines Devoted to Each Type of Pedagogical Move. 43
VII Distribution of Move Types According to Source. 44
VIII Relative Frequency of Most Common Transitions (Move Types).44
IX Per cent of Structuring Moves and Lines by Teachers and Students in Each Session of Each Class. 45
X Per cent of Soliciting Moves and Lines by Teachers and Students in Each Session of Each Class. 46
XI Per cent of Responding Moves and Lines by Teachers and Students in Each Session of Each Class. 47
XII Per cent of Reacting Moves and Lines by Teachers and Students in Each Session of Each Class. 48
XIII Comparison of Classes Three and Five According to Source of Pedagogical Move Types. 49
XIV Comparison of Patterns of Pedagogical Moves in Junior High School Mathematics Classes and Senior High School Social Studies Classes. Per cent of All Moves and All Lines Devoted to Each Move Type by Teachers and Students. 51
XV Per cent of All Moves and Lines Devoted to Major Content Topics in All Classes Combined by Session. 54
XVI Per cent of Moves and Lines Devoted to Each Type of Mathematical Activity in All Classes Combined by Session. 56

XVII Per cent of Moves and Lines Devoted to Each Mathematical Activity in Each Class. 57
XVIII Mathematical Code Cycles Per Session and Lines Per Cycle in Each Session of Each Class and in All Classes Combined. 58
XIX Per cent of Moves Initiating and Closing Mathematical Code Cycles by Teachers and Students in Each Class and for All Classes Combined. 60
XX Transition Frequencies for Mathematical Activity Within Content Cycles. 62
XXI Transition Frequencies for Mathematical Activity Focus Within Content Cycles. 62
XXII Transition Frequencies Within Content Cycles—Mathematical Activity and Focus. 64
XXIII Subordinate Cycles as a Portion of Total Activity in Each Class and for All Classes Combined. 65
XXIV Per cent of Subordinate Content and Activity Cycles Initiated and Closed by Teachers and Students in Each Class and for All Classes Combined. 66
XXV Per cent of Lines and Moves Devoted to Each Mathematical Activity in Subordinate Cycles in Each Class and for All Classes Combined. 68
XXVI Per cent of Logical Operations by Teachers and Students in Each Class and in All Classes Combined. 69

1

PURPOSE AND DESIGN OF THE RESEARCH

SOURCES OF THEORY ON TEACHING

Although systematic observation of classroom activity has only recently become an important concern of educational research, Jackson,[1] Travers,[2] and B. O. Smith[3] have argued forcefully that empirical studies of "the way teaching is" are a prerequisite for constructing a valid theory of instruction. In mathematical education Robert Davis has concerned himself with contributing to a theory of instruction. Faced with the problem of explaining what it is that makes some teachers more successful than others with his Madison Project classes, Davis said:

> What we seek. . . is a general description of what goes on in the classroom and in the school, from which we can begin to identify those variables which appear to be most decisive in determining success or failure, in the long run, for our program of "informal exploratory experiences."[4]

This alone is formidable support for descriptive-analytic studies of interaction in mathematics classes. However, by providing specific answers to the following questions, a still clearer rationale for such a study emerges:

1. Why is psychological learning theory an insufficient basis for theory of instruction?

2. Why are general teaching methods (such as the project and discovery methods) inadequate as theories of instructional behavior?
3. What justification is there for research on teaching which does not manipulate teacher variables and investigate the correlated changes in outcomes such as student achievement and attitudes?

The answers to these questions provide an insight into the purpose of this study and a rationale for its design.

Learning Theory and Teaching Theory

Current theory on the learning of mathematics fails as theory on teaching for two reasons. First, present learning theory is not a comprehensive set of generally accepted principles describing all or even most types of human learning. As Travers has observed:

> The global theories of learning of the thirties have come to be regarded by professional psychologists, not as global theories, but as more limited theoretical formulations.[5]

The investigations of Piaget and his disciples—while producing valuable information about the formation of such primitive concepts as the varieties of conservation—have shown that our understanding of the complex processes underlying human learning is just beginning. This is particularly true of the higher order concepts essential in mathematics. Concept formation is currently an active area of research in mathematical education. But even if such research produces a comprehensive theory of learning, a teaching theory will not necessarily follow as a routine corollary.

Learning is a change that occurs in the learner as a result of some experience; teaching is an activity intended to bring about learning. Thus, although learning theory has definite implications for teaching, these implications must be translated from statements about how learning takes place into recommendations for teaching behavior.

Jerome Bruner, a psychologist who has worked at this translation problem, cautions those who would infer a teaching theory from a learning one. The theory of programmed instruction, which claims that students should be taught in a sequence of very short steps, is derived from a psychological theory which says that learning occurs in small increments. Bruner states, however, that

> Nowhere in the evidence upon which such an instructional theory is based . . . is there anything that says that simply because learning takes place in small steps the environment should be arranged in small steps.[6]

Procedures for setting the learning environment are part of a theory of instruction. Thus, learning theory influences but does not determine the selection of classroom activities in teaching.

Learning theory fails as a direct source of teaching theory primarily because learning and teaching are fundamentally different kinds of activity, but also because it is an incomplete theory in itself.

General Teaching Methods

Despite the failure of psychology to provide theories of instruction (with the possible exception of programmed instruction), education has not gone without prescriptions of teaching methodology. Methods of instruction have been proposed which are derived from philosophy, logic, scientific method, and even psychotherapy. Project, laboratory, lecture, and discovery methods, along with many others, are general descriptions of the way teaching should proceed. (The discovery method, in particular, has recently attracted widespread interest in mathematical education.)

These teaching methods have commonly been developed by a combination of logical deduction and personal intuition. However, much research has been conducted comparing the effectiveness of different methods. After reviewing this research, G. Max Wingo stated:

> Reports of this research reveal the inconclusive nature of much of the evidence obtained and in many cases demonstrate that the evidence taken as a whole, is internally contradictory.[7]

This conclusion is easy to understand in light of his further contention that

> Patterns of instruction or general methodologies are difficult to categorize, and even when they are given a name, the patterns as they actually work out in various types of situations may differ greatly.[8]

An experiment, then, which attempts to manipulate ill-defined or unmeasurable variables is bound to yield inconsistent or unreliable results.

The disappointing findings of research on teaching methods do not rule out such prescriptions as a potential source of instructional theory. Instead, they suggest that the methods must be defined in behavioral terms which can be reliably observed and precisely measured. Careful study of classrooms in action is an important prerequisite to building such definitions.

Correlates of Teaching Effectiveness

The most common objective of research on teaching has been to identify the personal characteristics which typify effective teachers. The usual design of such research—in fact, the classic experimental design—involves manipulating specified teacher variables and investigating the correlation between this variation and student outcomes. A wide variety of teacher characteristics have been proposed and tested as determinants of teaching success, among them intelligence, education, knowledge of subject, experience, attitudes about students, sex, and age. While it would be foolish to contend that none of these attributes correlate with teaching success, few striking or surprising relationships have been found. The evidence from correlational studies does not give a comprehensive or consistent characterization of "the good teacher."

Opening a 1959 conference on "Psychological Problems and Research in Mathematics Training," C. R. Carpenter stated:

> Predominantly and with distressing frequency, research on single variables of methods of instruction and on conditions of learning has yielded "no significant differences."[9]

Carpenter's opinion is strongly supported by Jackson,[10] Biddle,[11] and others.

There is one obvious explanation for the failure to find predictors of teaching success: there must be certain critical teacher variables which have not yet been identified or tested.

A two-year Minnesota investigation of *Characteristics of Mathematics Teachers That Affect Students' Learning* found no significant correlation between teaching effectiveness and the usual teacher variables—experience, quantity and quality of mathematical education, and professional activity.[12] However, probing a new class of teacher variables, this study did discover a relationship between student learning and the productive thinking ability of their teachers.

> The most effective teachers produced a greater variety of ideas about indications of success and failure in their teaching, hypothesized a greater variety of causes of success and failure, and offered a greater variety of alternative ways of teaching mathematical concepts than did their less effective peers.[13]

These results are encouraging. However, even if similar studies identify other important teacher characteristic variables as a source of theory on the teaching process, such findings are of limited practical value. Teacher attributes correlated with competence are primarily useful as predictive measures in the selection of those who will teach. Smith has pointed out that "such factors as personality traits, intelligence, and knowledge of instructional content are static elements of teacher behavior."[14] Measures of these teacher variables indicate nothing about the way a teacher behaves in the classroom. To study teaching without looking at classroom activity is to overlook the dynamic aspect of the teaching process.

This is the conclusion reached by Muriel Wright, who argues that study of teaching by direct observation of class activity will achieve "simultaneous consideration of the subject matter taught and the method of its development—two interdependent facets of the lesson."[15]

A theory of teaching is a set of hypothesized relations between elements of the teaching–learning process. As has been shown, attempts to confirm theories about teacher characteristics and student outcomes have frequently shown that the hypothesized relations either do not exist or account for but a small portion of outcome variability. Therefore, research must begin to test new relations between the previously studied variables or locate entirely new clusters of variables in mathematics teaching. The study reported in the following chapters was, in part, an attempt to identify new variables in the classroom behavior of mathematics students and teachers. The obvious hope is that since this behavior is a central aspect of teaching, such variables should provide a fruitful point of departure for renewed study of mathematics teaching.

SIGNIFICANT RELATED RESEARCH

Failing in their search for reliable predictors of teaching success among static teacher variables, recent investigators have turned to the classroom in pursuit of behaviors which characterize effective teaching.[16] As David Ryans has pointed out:

> Teacher evaluation, or the judgment of teacher effectiveness, can be properly and successfully accomplished only when it is based upon reliable knowledge of the essential behaviors involved. . . . Thus, it is appropriate that much of the research being conducted today is concerned with the identification of the behaviors of teachers and with their descriptions, rather than proceeding in value considerations.[17]

Various investigators are studying teaching from the view point of linguistics, logic, communication theory, and sociology. But all of these studies have one common goal: the discovery of new—

and hopefully significant—teacher variables through the systematic study of classroom activity.

Recent and current approaches to the study of classroom behavior are of two basic types: those which focus on the interpersonal aspects of classroom interaction and those which focus on the cognitive aspects. A few have attempted to analyze both affective and cognitive dimensions simultaneously.

As Nathaniel Gage has observed:

> Many if not most of the best known programs of research on teaching have been aimed at social and emotional aspects of how teachers behave and how pupils respond and develop.[18]

These investigations, with basically psychological and sociological background, have explored teacher behavior in the hope of identifying those activities which influence classroom climate and, as a consequence, learning.

Typical, and perhaps of most importance, among these socio-psychological studies of the teaching is the series of investigations begun in 1955 by Ned Flanders,[19] who developed a system of *interaction analysis* that "provides an explicit procedure for quantifying direct and indirect [teacher] influence that is closely related to the teacher behavior identified by research on classroom climate."[20]

Using interaction analysis, Flanders tested a variety of conjectures concerning the relationship between classroom climate (assumed to be determined largely by the teacher) and student achievement and attitudes. One of his tests involved a study of 16 mathematics and 16 social studies classes to examine the following hypotheses:

1. Indirect teacher influence increases student learning when a student's perception of the goal is confused and ambiguous.
2. Direct teacher influence increases learning when a student's perception of the goal is clear and acceptable.
3. Direct teacher influence restricts learning when a student's perception of the goal is ambiguous.[21]

All three hypotheses were confirmed. Successful teachers (measured by achievement and attitudes of their students) were those who consistently exerted more indirect influence than direct influence. These successful teachers also showed a tendency to move from extreme indirect influence at the beginning of a unit of study, when goals were more ambiguous, to more direct influence as goals of the unit became clearer.

Several other studies using interaction analysis have confirmed

Flanders' contention that teachers who use more indirect influence than direct influence produce better student achievement and attitudes. Flanders, Edmund Amidon and Elizabeth Hunter began translating their findings about teacher influence into prescriptions for teacher improvement.[22]

The objective of research in interaction analysis, and of similar studies of teacher influence, is better understanding of the way controlled verbal communication can be used as a social force in classroom management. The research reported in this book starts from the assumption that it is equally important for teachers to understand the way controlled verbal communication can be used to structure the cognitive aspects of classroom interaction.

> In research on teaching for cognitive objectives we have had relatively little of the descriptive, analytical, theoretical, experimental, or correlational work that can be found in relative abundance in research on the social and emotional phenomena found in classrooms.[23]

Among studies that have dealt primarily with the intellectual dimension of classroom communication, two general studies are of particular significance: Smith's *A Study of the Logic of Teaching* and Arno Bellack's *The Language of the Classroom.*[24]

Analyzing transcripts of English, social studies, science, and mathematics classes, grades 9 through 12, Smith found that classroom verbal activity consists of *episodes*—"one or more exchanges which comprise a completed verbal transaction between two or more speakers"—and *monologs*—"solo performances of a speaker addressing the group."[25] Moreover, these episodes and monologs exhibit recurrent identifiable structures which can be analyzed in logical terms. Thirteen logical operations were found to occur in teaching with *describing, designating,* and *explaining* most common. Although relative frequency of occurrence of the logical operations varied from class to class, no definite conclusions could be drawn due to an inability to isolate the teacher from content variables. Furthermore, tapes from two of the three mathematics classes could not be analyzed because of the technical nature of the discussion and excessive references to symbols written on the chalkboard.

The second phase of the Smith study focused on three different units of classroom verbal interaction: moves, strategies, and ventures.[26] Analyzing the transcripts used in the first study, Smith and associates identified nine types of ventures—within each, many types of moves—and four types of strategies. Development of these concepts is still in progress, but they offer interesting new ways to conceptualize verbal instructional behavior.

The second investigation which has made a major contribution to the study of cognitive classroom interaction is that of Arno Bellack at Teachers College.[27] Examination of transcripts from 15 high school social studies classes suggested that classroom verbal communication could be viewed as a kind of language game. Students and teachers are the players in this game, and communication of various kinds of meaning is the object. Bellack identified four basic moves used in the classroom language game, and discovered rules which apparently govern the source, cognitive objective, frequency, and ordering of these moves.

The studies of Smith, Bellack, and others offer various ways of conceptualizing cognitive aspects of classroom verbal communication. Smith found that tapes of mathematics classes were difficult to code because of the technical and symbolic nature of the discussion. However, he assumes that logical analysis of teaching is independent of subject matter. Bellack hypothesizes that, with suitable modification in substantive categories, the language game model might prove fruitful in analyzing discussions in English, science, and mathematics classes.

It seems reasonable to expect that the methods and substantive structure of mathematics place unique restrictions on the pattern of verbal communication in mathematics classes. The present study was, in part, an attempt to measure the influence of mathematical content on patterns of communication. The most significant previous work in this direction was that of Wright.

In developing an instrument for studying verbal behavior in mathematics classes, Wright chose categories suggested by objectives of mathematics teaching. As first conceived, the categories fit into three major frames:

FRAME A. Conscious developing (by teacher) or use (by teacher or pupil) of ability to think. . . .

FRAME B. Conscious developing (by teacher) or demonstration (by teacher or pupil) of appreciation of mathematics. . . .

FRAME C. Conscious fostering (by teacher) or demonstration (by teacher or pupil) of an attitude of curiosity and initiative. . . . [28]

Observers were to study verbal activity in the classroom for a 15–second interval and then (during the next 15–second interval which was not observed) classify the behavior in one of the sub-categories of each frame. This system was used to analyze interaction in 12 algebra classes with the primary objective to test the reliability and validity of the instrument.

In 1961 Wright and Virginia Proctor revised the instrument by eliminating the *appreciation* frame and introducing a frame covering mathematical content.[29] This revised system of analysis was used to assess simultaneous development of subject matter, cognitive process, and attitudinal factors in 12 classes. Counts were made of entries in each category and in certain combinations of categories in an attempt to discriminate activity in the various classes and styles of teaching.

The objective of the Wright system of analysis—simultaneous consideration of mathematical content, cognitive process, and affective factors—is important for the study of classroom interaction in mathematics teaching. However, there are three features of the instrument which can and should be improved.

First, the time sampling method—coding 15-second blocks with the intervention of 15-second unobserved blocks—measures only half the classroom activity. Even allowing for the fact that behaviors fall at random into coded or uncoded intervals, the continuity of classroom discourse is not adequately preserved by such coding procedures.

Second, in describing individual intervals, the observer has only 15 seconds to choose codes from among 50 possible categories in three frames. Although tests of reliability indicate that it is possible to attain proficiency in such complex rapid fire coding, such reliability can be gained only by sacrificing sensitivity in the instrument.

Third, the categories within the *content frame* do not constitute a unified description of the substance and activity of mathematics. The category of Fundamentals (defined as the examination of previously developed mathematics) has two sub-categories: structure and techniques. However, the category of Relations (defined as developing new properties in a mathematical system) includes three quite different types of categories: induction, deduction, and statement. It would seem natural for the relations group to include sub-categories paralleling structure and techniques. Furthermore, induction, deduction, and statement are more properly considered processes than content.

Descriptive-analytic studies of classroom verbal interaction have two common objectives: to conceptualize the classroom communication phenomenon, and to develop observational schemes for describing classroom activity in terms of this conceptualization. The natural question is: What does this work contribute to the improvement or evaluation of teaching? What is the payoff from describing actual teaching behavior?

Making use of concepts developed by Smith (move, strategy, venture, etc.), Kenneth Henderson has devised several experimental studies to determine which of several possible strategies are most effective for teaching mathematical concepts.[30] Several teaching treatments are constructed by combining possible moves in various ways. Then student achievement is compared as a measure of effectiveness of the strategies.

Since 1961 the Minnesota National Laboratory has been evaluating experimental mathematics programs in the secondary schools of a five–state area. As part of this evaluation observational systems were used to answer the following questions:

1. Is there a pattern of involvement behaviors for an experimental and for a control classroom? Do the experimental curricula draw the pupils more into all aspects of learning?
2. Is there a pattern of content behaviors for an experimental and for a control classroom? Do the experimental curricula increase attention to theoretical aspects of mathematics? Do they increase emphasis of development of new relationships in mathematical structure? Do they neglect practice in specific problem solving skills?[31]

A modified interaction analysis was used to examine the first question, and one using the content frame of the Wright-Proctor instrument the second.

The hypothesis concerning greater student involvement in experimental programs was not confirmed. Increased attention to structural aspects of mathematics was observed in the experimental programs without decline in time devoted to skills. But the significance of this research for the present discussion is the fact that two conceptualizations and observational techniques were useful in the experimental evaluation of teaching. It is hoped that the research reported in later chapters also contributes to this task.

RESEARCH DESIGN

Systematic study of classroom verbal communication is a promising new direction in research on teaching. Knowledge of the characteristic classroom behaviors of teachers and students will provide basic concepts to be organized and interrelated by theory on instruction. Furthermore, observational instruments designed to describe patterns of verbal interaction will be useful tools in experimental studies where teacher behavior is a central variable.

The structure of mathematics can be expected to exert a unique influence on patterns of verbal behavior in mathematics classes. Therefore, this study undertook two main tasks: (1) To develop an instrument which describes the pedagogically and mathematically significant components of teacher–student verbal communication. (2) To use this instrument to describe patterns of verbal communication in the five mathematics classes participating in the Secondary School Mathematics Curriculum Improvement Study (SSMCIS).

Any attempt to study classroom activity by systematic observation confronts two basic problems. First, it is necessary to make an accurate record of the observed behavior. Second, this record must be partitioned into units which can be described in meaningful terms. Objectives of data analysis influence the selection of observational techniques, and conversely, the quality of behavioral records places limitations on the reliability of inferences to be drawn from the data. Because formulating the system of analysis was itself an important part of the study, complete and accurate observational records were required before specific data analysis could be planned. These records were obtained in the following way:

1. When each of the five classes began study of the chapter "Multiplication of Integers" in the SSMCIS textbook, four consecutive meetings were (audio) tape–recorded. The recorder and a single microphone were placed unattended in the middle of the class.
2. An observer, present at each recorded class meeting, made detailed notes, including descriptions of all writing done at the chalkboard and special non-verbal action deemed necessary to complete the tape–recorded picture of the class meeting.
3. Each tape–recording was transcribed and coordinated with the observer's notes.

Comparison of the augmented tapescripts and the observer's impressions of classroom activity confirmed the hope that the above procedures would result in accurate behavioral records. Students' and teachers' voices were clearly recorded; the observer's notes contained the significant written communications; and the presence of a tape–recorder did not disrupt the normal pattern of class activity. The permanence and completeness of the tapescript records permitted deliberate and repeated examination of each class meeting.

Accurate behavioral records were the first goal of the study. But description and comparison of these records were more important and more difficult tasks. Using ideas from communication theory, logic, and linguistics and concepts developed in earlier studies of Smith, Bellack, and Wright, a system was devised analyzing the pedagogical function, duration, content, mathematical activity, and logical purpose of each utterance in classroom discourse about mathematics. Transcripts of the tape–recorded class meetings were coded using this system, and the coded protocols were analyzed to determine patterns of verbal communication in the selected classes.

The system of analyzing protocols, described in detail in Chapter II, is summarized below.

The System of Analyzing Protocols

1. Speaker — teacher (T) or student (S).
2. Type of Pedagogical Move[32] — role of the utterance in shaping communication of ideas.
   a. Structuring (STR) — moves which set context for subsequent discussion by focusing on a topic or activity.
   b. Soliciting (SOL) — moves which elicit information or action.
   c. Responding (RES) — moves which fulfill expectation of soliciting moves.
   d. Reacting (REA) — moves which modify, expand, or rate the content of prior moves.
3. Duration — number of lines in the type–written transcription of class recordings.
4. Content — major mathematical topic under consideration.
5. Mathematical Activity — role of the move in developing, examining, or applying mathematical systems.
   a. Developing (D) — defining and building properties of mathematical systems.
   b. Examining (E), Recalling (R), Illustrating (I), or Comparing (C) — studying structure of previously developed systems.
   c. Applying Mathematical Systems — used to solve mathematical (AM) or non-mathematical (AN) problems. Mathematical activity is further coded according to the emphasis on Elements, Operations, Mappings, Relations, Logic, or Theorems in a system.
6. Logical Process — cognitive activity involved in dealing with mathematical content.
   a. Analytic Process (Aly) — statements about meaning and usage of language and symbolism.
      (1) Defining (Def)
      (2) Interpreting (Int)
   b. Factual Process (Fap) — statements which give or solicit information about a mathematical system.
      (1) Fact-statement (Fac)
      (2) Conditional inferring (Inf)

c. Evaluative Process (Val) — statements about the truth or appropriateness of a previous remark.

d. Justifying Process (Jus) — statements which support a previous remark.
   (1) Deducation (Ded)
   (2) Induction (Ind)
   (3) Opinion or Authority (Opn)

2

THE SYSTEM FOR ANALYZING BEHAVIORAL RECORDS

Various naive analyses of classroom tapescripts will yield only partial profiles of the verbal communication in recorded meetings. Such analyses might include measures of relative frequency and duration of student and teacher talk, frequency of occurrence of various grammatical forms, or emphasis placed on mathematical as opposed to non–mathematical subject matter.

Each of these measures provides some insight into a particular aspect of classroom discourse. However, verbal communication is a multi–dimensional activity—a complex ordered process in which content is shaped into linguistic form by communication sources and interpreted by receivers of the speech. It seems reasonable that analysis of classroom verbal activity should begin by dividing the discourse into communication units, each of which can be described according to source, target, duration, substantive content, linguistic form, logical function, and other significant dimensions of the communication process. Of course, any such segmentation into descriptive units should be done in a manner which retains as much as possible of the vital sequential nature of the discussion.

THE CLASSROOM AS A COMMUNICATION SYSTEM

There are several broad concepts of verbal communication which could guide analysis of classroom discourse. One promising approach is suggested by Shannon and Weaver in *The Mathematical Theory of Communication.* In this book the authors propose a five–

part paradigm for any communication process. A message is originated in the *source* (in classrooms, teacher or student) and coded in a form that can be sent by the *transmitter.* The transmission is carried over a *channel* to a *receiver* and then decoded for understanding by the *destination.*[1]

Source	→	Transmitter	→	Channel	→	Receiver	→	Destination
(T or S)		(voice)		(air)		(ear)		(T or S)

The present study focuses on coding process. In particular, it examines the function of a semantic coding unit located between the source and the transmitter.

Source → Semantic Coding Unit → Transmitter

This semantic coder is the device for translating meaning into appropriate linguistic form. One goal of the data analysis is to identify and compare coding procedures used in the observed classroom communication networks. Ultimately the study of successful classroom communication should identify those coding procedures which appear to lead to optimal communication of the desired ideas.

A second possible framework for studying verbal communication is that which Bellack and his associates developed from ideas of Wittgenstein. In *The Language of the Classroom,* classroom discourse is viewed as a kind of *language game.* Students and teachers are the players in the contest whose object is the transmission of meaning by various kinds of linguistic moves. The Bellack study found that the classroom language game has apparent rules governing the source and sequence of the characteristic moves.[2]

Within this concept, the present study attempts to identify moves which are characteristic of mathematics classes as well as the way these moves are put together in playing the verbal communication game.

While such a broad view of verbal communication is helpful for organizing observations into a coherent picture of classroom activity, the conceptual framework itself is of little help in describing specific communication units.

It is tempting to assume that the formal structures inherent in language and mathematics provide ready-made categories encompassing all possible forms of communication utterances. Unfortunately, spoken language is not a precise, logical presentation of ideas. Some sentences containing distinct thoughts are freely run together; others are left half completed with the listener to infer the conclusion from context; still other utterances do not have as a purpose either direct statement or question of fact.

There are also pedagogical activities occurring in the mathematics classroom which have no close counterpart in formal mathematical activity: review of previous learning, drill to enhance computational skills, and a variety of classroom organizational procedures, to name a few. Furthermore, when basic mathematical processes such as definition and proof do occur in the classroom, they do so in far from ideal form. For example, the following excerpt from a tapescript shows a student defining "isomorphism":

S: I think it's isomorphism.
T: What is that?
S: They both got the same amount, and you could change them around and it's the same thing.

In this brief interchange the student claims that a situation involves an isomorphism. The teacher asks him to define "isomorphism," and the student replies with a highly intuitive explanation, not a precise mathematical definition.

Linguistic form, logic, or mathematical processes can form the basis of a system for analyzing classroom discourse. However, if the system is to provide a truly valid description of the nature and function of verbal utterances, it must add to these other theoretical behaviors which actually occur in typical classrooms. The following descriptive system consists of categories hypothesized as significant in communication (and subsequently found to actually occur in consistently recognizable form) and categories suggested by examination of the tapescripts. The discussion utilizes the vocabulary of the language game model of classroom verbal activity. However, the results, with appropriate translation, can be stated equally well using the communication system model.

THE UNIT OF ANALYSIS

The first major problem in developing a method of analyzing records of classroom discourse is defining a unit of analysis—a segment of the discussion which can be identified reliably and described in meaningful terms. One highly reliable unit would be a fixed time interval of perhaps five seconds. However, classroom discourse does not naturally break on such a rigid schedule. A more meaningful partition might be according to kinds of classroom activity: homework, new material, testing, seat work, etc. But these analytic units give only a gross picture of the actual verbal interaction by which teaching progresses.

It is conceivable that one might define several units of analysis—a time unit, a content unit, a grammatical unit, and a class activity unit—or a single unit to be determined by a composite of criteria. Preliminary investigation of the tapescripts showed that any large or composite unit is very difficult to identify reliably. However, this investigation also showed that the Bellack pedagogical moves,[3] originally developed in order to describe verbal activity in social studies classes, were also readily identifiable in the protocols of mathematics classes.

With the move as a basic unit, it is possible to construct a variety of larger units as sequences of moves. Therefore, the following coding procedure was adopted:

1. Each tapescript was partitioned into a sequence of moves: structuring, soliciting, responding, reacting.
2. Each move was then described according to source, length, content, mathematical activity, and logical process.

CODING MOVES

The pedagogical move used by Bellack is an uninterrupted utterance or partial utterance of a single speaker which serves the pedagogical purpose of structuring the discourse, soliciting information or action, responding to a solicitation, or reacting to a prior move. Each move can be classified further by coding according to the other criteria mentioned above. This coding is described in detail and illustrated in the following sections.

Pedagogical Purpose

The pedagogical purpose of each move is determined according to the definitions given by Bellack.

> *Structuring* (STR). Structuring moves serve the function of setting the context for subsequent behavior by (1) launching or halting interactions between teacher and pupils, and (2) indicating the nature of the interaction in terms of the dimensions of time, agent, activity, topic and cognitive process, regulations, reasons, and instructional aids. A structuring move may set the context for the entire classroom game or a part of the game.[4]

For example:

T/STR: Let's turn now to multiplication.
T/STR: What we would like to do here is to start with a and b. We must

find rules for assigning the product of a and b in four cases.
S/STR: I want to ask a question about something else.

> *Soliciting* (SOL). Moves in this category are intended to elicit (a) an active verbal response on the part of the persons addressed; (b) a cognitive response. . . or (c) a physical response.[5]

Examples:

T/SOL: If we were adding two directed numbers, both of them going to the left, what would our result be?
T/SOL: What is the sum, then, of -83 and -27, Allan?
S/SOL: Why is a negative times a negative always a positive?

> *Responding* (RES). Responding moves bear a reciprocal relationship to soliciting moves and occur only in relation to them. Their pedagogical function is to fulfill the expectation of soliciting moves.[6]

Examples:

T/SOL: What is zero called?
S/RES: Identity.
T/SOL: If we add those two integers, what's our result?
S/RES: -5.
S/SOL: What is our assignment for Monday?
T/RES: No problems, just read.

> *Reacting* (REA). These moves are occasioned by a structuring, soliciting, responding, or a prior reacting move, but are not directly elicited by them. Pedagogically, these moves serve to modify (by clarifying, synthesizing, or expanding) and/or to rate (positively or negatively) what was said in the move(s) that occasioned them.[7]

For example:

T/REA: Yes, you could look at it that way.
S/REA: I don't think that what Susan said is right.
T/STR: The question was, why can't we multiply two positives and get a negative?
S/REA: No, it's two negatives to get a positive.

No significant modification of the definitions of these moves was necessary to obtain high reliability in coding the protocols of mathematics classes. However, more detailed instructions given the coders are presented in the Appendix: "Instructions for Coding."

Source

The source of each move was coded *student* (S) or *teacher* (T). While it would have been desirable to identify which particular student was speaking in S moves, the tape could not be reliably used for this information. In some places, where such knowledge was of particular significance for understanding the discussion, explanatory notes were added to the tapescript. For example:

T: But you (the speaker 3 moves above) said it was -5.

Duration

The length of each move was coded as the number of half–lines of typewritten copy covered by the transcription of that move. (One line is five inches of elite type.) No move was coded as less than one half–line. Partial lines beyond the first were counted if they covered more than half of the next unit. For example:

T: Ruth, do you know the distributive law of multiplication over addition? (T/SOL/3)
S: I'm not sure, but I think that it is $a(b+c) = ab + ac$. (S/RES/2)

When an utterance contained more than one move, line fragments were put together to determine the duration of each:

T: That's good./ Now let's turn our attention to the second set. (T/REA/1)/(T/STR/2)

Underlying this choice of duration of move measure was the assumption that length of time required to speak is closely related to the length of the corresponding transcription. Even if this is not the case, it seemed reasonable to assume that total word production is an appropriate measure of the duration of a communication utterance.

Content

Since the main interest of this study was the way mathematical ideas are communicated and developed in the classroom, the content of a move was described in detail only if the move had meaning relevant to mathematical ideas under consideration. Moves concerned solely with classroom procedural matters were coded "P." For example:

T: Let's pass the papers in now. (T/SOL/1/P)
S: Would you open the window? (S/SOL/1/P)

When a move had mathematical content, it was categorized in one of the following classes determined from a content analysis of the SSMCIS textbook chapter being studied (Multiplication of Integers):[8]

Integers: Z —general references to the integers, not more specifically codable.
Zm —multiplication of integers.
Za —addition of integers.
Zs —subtraction of integers.
Zp —general references to the integers as ordered pairs of whole numbers.
Zpm —multiplication of integers expressed as ordered pairs of whole numbers.
Zpa —addition of integers expressed as ordered pairs of whole numbers.
Zv —absolute value of integers.
Zo —ordering of the integers.

Finite systems: Zf —references to modular arithmetic.

Whole numbers: W —general references to the whole numbers, not more specifically codable.
Wm —multiplication of whole numbers.
Wa —addition of whole numbers.
Ws —subtraction of whole numbers.
Wd —division of whole numbers.

Directed numbers: $\overline{W}$ —general references to the directed numbers.
$\overline{W}$a —addition of directed numbers.

Dilations: D —general references to the set of dilations of a line under composition.
Dc —composition of dilations.
Dv —magnitude of dilations.
Dr —reflections.

When several systems were being discussed at the same time (as in consideration of isomorphism), the content was coded with two letters indicating the major content classes. For example, "ZW" refers to the integers and the whole numbers being studied simultaneously. "W$\overline{W}$" refers to the whole numbers and the directed numbers being studied simultaneously.

The above categories of content meaning exhaust the major topics contained in the chapter that was studied during the tape-recorded class meetings. Other relevant mathematical or non-mathematical substantive references were coded "Rel," non-relevant topics "Nrl."

As a general rule, when a move fitted two or more content classifications, the more specific description was chosen. For example:

T: What is the sum of the integers given by the pairs (2, 3) and (7,1)?

was coded Zpa because this gave a more specific classification than either Z or Zp, both of which were appropriate. In some cases conflicts could not be settled in this way. For instance:

T: If $a < b$, then $a + c < b + c$,

involved both addition and ordering of the integers. Neither Za nor Zo could be considered more specific. Therefore, the move had to be classified simply Z.

Mathematical Activity

Describing moves according to source, pedagogical purpose, duration, and substantive content yields an informative sequential summary of class discussion. This summarizing process is illustrated by the following segment of a coded transcript.

T: We have been talking about multiplication of integers. We have a pair of integers and to this pair we wish to associate by multiplication a single integer c. We also want this assignment to agree with what has been done in earlier number systems./ (T/STR/11/Zm) What is one property of (W,·)* that we want to hold also in (Z,·)? (T/SOL/3/Wm)
S: Commutative property. (S/RES/1/Wm)
T: All right./ (T/REA/1/Wm) What is another? (T/SOL/1/Wm)

The brief code following each move above shows that the teacher set the context for a discussion by a lengthy statement about multiplication of integers and then asked a brief question about whole number multiplication. A student responded to this solicitation with a short statement about (W,·); the teacher reacted to this response, and then went on to ask another question on the same topic.

The data obtained from this type of coding suggest many interesting questions. For example, in the observed mathematics classes:

1. What is the ratio of student and teacher moves in terms of frequency and of duration?
2. What are the relative frequencies of occurrence of structuring, soliciting, responding, and reacting moves? Are these relative frequencies constant from class to class or from one day to the next in a single class?

*"(W,·)" and "(Z,·)" refer to the multiplicative monoids of whole numbers and integers, respectively.

3. Do all classes concentrate attention on the same substantive topics?
4. What are the transition probabilities from one type of pedagogical move to another?
5. How do relative frequency, duration, and sequencing of the pedagogical moves compare with those in social studies or literature classes?

These and other questions are considered in Chapter III, which contains results of the protocol analysis. But student–teacher discussion of mathematics is not simply a sequence of moves with identifiable topic and pedagogical character; it is a purposeful development of ideas. It is reasonable to suspect that the structure of mathematical ideas has a unique influence on the pattern of discussion in mathematics classes. To measure this influence, a system of analysis must go beyond the four–step procedure outlined so far, and must describe the mathematical activity involved in the discussion.

Any arbitrary imposition of hypothesized mathematical behavior categories on classroom discourse is likely to give a misleading picture of the content and purpose of communication moves—completely missing one of the main objectives of systematic study of classroom activity. In fact, several analytic systems of armchair derivation were quickly dropped after attempts to apply them to actual class recordings failed. Things happen in the classroom which have no counterpart in formal mathematics. Where classroom acts do have mathematical structure, the connection is often hard to recognize.

Despite the difficulty and danger of applying a conjectured structure to classroom discourse, it is impossible to approach analysis of tapescripts without some pre–conceptions of the form the discourse *might* take. The categories of the present system of analyzing mathematical activity evolved from continuous interplay between conjectured structures and the record contained in the tapes.

The conceptual approach that ultimately led to the system described below is based on the assumption that student–teacher mathematical activity is the activity of people doing *mathematics*. Thus, one would expect to find verbal behavior in the classroom which parallels the behavior of a person doing mathematics. As a starting point, mathematical activity was assumed to have one of two main objectives: (1) *developing* mathematical systems (coded D), or (2) *applying* mathematical systems to mathematical (coded AM) or non–mathematical (coded AN) problems.

Inspection of the protocols quickly showed that there were other activities used for pedagogical purposes in the classroom which could be described accurately either as developing or applying mathematical systems. Therefore a third major category was devised to encompass *examination* of previously developed mathematical systems. Examination was assumed to mean *recall* (coded R), *illustration* (coded I), or *comparison* (coded C) of properties of mathematical systems—recall and illustration being activities with primarily pedagogical objectives, and comparison including such creative mathematical activity as the search for isomorphism between algebraic systems.

Within these broad activity areas it seemed important to identify the particular aspect of a mathematical system being considered. A formal mathematical system consists of:

1. An underlying language.
2. A deductive logic system.
3. A vocabulary of undefined words.
4. A set of axioms—statements about the undefined words.
5. Theorems—statements about the undefined words that can be demonstrated.[9]

However, this characterization of mathematics emphasizes the philosophically primitive elements of mathematics, not the working concepts such as sets, operations, relations, and functions used by mathematicians in most of their work. For this reason, the following outline was used in this study to describe the aspect of a mathematical system being developed, recalled, illustrated, or compared:

- 1.0 Objects of the system (primitive or defined)
    - 1.1 Specify
    - 1.2 Describe properties of (axioms)
- 2.0 Binary operations
    - 2.1 Definition or statement of computational procedure
    - 2.2 Properties
        - 2.21 Computational facts
        - 2.22 General (quantified) properties
- 3.0 Mappings (not including binary operations)
    - 3.1 Definition
    - 3.2 Properties
        - 3.21 Representation or computation of images
        - 3.22 General properties
- 4.0 Relations (not including functional relations)

4.1 Definition
4.2 Properties
4.21 Representation or recognition of relation between a specific pair of objects
4.22 General properties
5.0 Logic—nature, proper form, and fnction of definitions, axioms, theorems, proof, etc.
6.0 Theorems—statements of general or specific nature involving objects, operations, mappings, and relations of a system (when not codable more accurately as 1, 2, 3, or 4)

Using this scheme, the mathematical activity of each move can be coded according to type of activity (D, R, I, C, AM, AN), and the focus of that activity (objects, operations, mappings, relation, logic, theorems). These coding procedures are defined more precisely and illustrated in the following:

A move was coded *developing mathematical systems* (D) if its basic objective was to specify the elements of a system and/or establish properties of operations, relations, or mappings in a system.

As with all other mathematical activity categories, it is difficult, if not impossible, to classify an isolated move as developing mathematical systems. Therefore, coding was done according to the function the utterance performed as part of continuous discourse. For example, the move

S: What does it mean?

does not, out of context, represent a question about the definition of a mapping. However, when taken as part of the following discussion, the function of the utterance is clear.

T: Now we're gonna go back into the realm of mappings. I'm gonna introduce a new type of mapping. A mapping which is related to a point which I shall call C. Under this mapping, C maps to itself. Suppose I take any other point on the line, say P. Any point. P can be here, P can be here, here; any point along this line. I am going to map P to a point P′ such that the line segment CP equals PP′. (T/STR/18/D/D3.1)
S: I don't . . . So what about it? (S/SOL/1/D/D3.1)
S: What does it mean? (S/SOL/1/D/D3.1)
S: First of all, what is CP? (S/SOL/1/D/D3.1)
T: C is the point from which I want to take all my mappings. C is a central point. C always maps to itself. (T/RES/5/D/D3.1)
S: Why that? (S/SOL/1/D/D3.1)

T: Any other point either to the right or to the left of C, say I take a point over here, this point Q, I look at the distance from C to Q and I go the same distance from Q and this would be? (T/SOL/8/D/D3.21)
S: Q′. (S/RES/1/D/D3.21)
T: Q′. / Notice, what relationship does Q have to Q′ and C, Bruce? (T/REA/1/D/D3.21)/(T/SOL/2/D/D3.21)
S: It's like the midpoint. (S/RES/1/D/D3.21)
T: It's like the midpoint! It is the midpoint. (T/REA/2/D/D3.21)

As the coding indicates, the teacher introduced a mapping of the set of points on a line. The discussion first focused on the definition of the mapping, then on computation of images.

A move was coded *comparison* (C) if its basic objective was to point out similarities and differences between mathematical systems. The preceding discussion continued into such activity, comparing dilations of a like with mappings of the integers:

S: Why does C map to itself? (S/SOL/1/D/D3.1)
S: It's zero. (S/RES/1/DZ/C1.0)
T: It's like a zero. (T/REA/1/DZ/C1.0)
S: Oh, you're doing it like the number line? (S/REA/2/DZ/C1.0)
T: It's like. . . . It resembles the number line very closely. (T/REA/2/DZ/C1.0)
S: If Q would be negative 1, then Q′ would be. . . (S/STR/2/DZ/C3.21)
T: What if Q was negative 2? (T/SOL/1/DZ/C3.21)
S: Then Q′ would be negative 4. (S/RES/1/DZ/C3.21)
T: Right. (T/REA/1/DZ/C3.21)
S: 2n (S/STR/1/DZ/C3.21)
T: Ahh, what was that mapping? (T/SOL/1/DZ/C3.22)
S: 2n (S/RES/1/DZ/C3.22)
T: n maps to 2n. Very nice. (T/REA/1/DZ/C3.22)

A move was coded *recall* (R) if its basic objective was to review properties of previously developed mathematical systems. Although serving primarily pedagogical purposes, recall is often an important prerequisite for building new systems. For example:

T: All right, before we go into the multiplication of integers, I'd like to review some of the properties of the multiplication of whole numbers. I would like to relate these properties of multiplication to the integers. / What properties of multiplication did the whole numbers have? Take the set W and we would like to work with multiplication. What was one property which existed under this situation, Jamie? (T/STR/9/Wm/R2.22) / (T/SOL/8/Wm/R2.22)
S: The commutative. (S/RES/1/Wm/R2.22)

T: Can somebody express a generalization for the commutative property? Mark? (T/SOL/3/Wm/R2.22)
S: *a* times *b* equals *c*. (S/RES/1/Wm/R2.22)
S: and *b* times *a* equals *c* (S/RES-M2/1/Wm/R2.22)
S: *a* times *b* equals *b* times *a*. (S/RES-M3/1/Wm/R2.22)
T: All right. What you were saying, *a* times *b* equals *c* and *b* times *a* equals *c* means *a* times *b* equals *b* times *a*. (T/REA/4/Wm/R2.22)

A move was coded *illustration* (I) if its main objective was to give an example of a general property in a mathematical system. For instance:

T: Can you give me an example of the integers having an additive inverse? All right, Jane? (T/SOL/4/Za/I2.22)
S: Positive one plus negative one. (S/RES/1/Za/I2.22)
T: All right, equals zero. / Diana, you give me one. (T/REA/Za/I2.22) / (T/SOL/1/Za/I2.22)
S: Positive 2 plus negative 2 equals zero. (S/RES/Za/I2.22)
T: Okay, fine. (T/REA/1/Za/I2.22)

Moves not specifically codable as R, C, or I, were labeled *examining* (E).

The coding of moves as *application* (AM or AN) raises difficulties because of various possible interpretations of the word "application." Ideally, application of a mathematical system is problem–solving activity involving three basic steps:

(1) Recognizing that a problem situation embodies the structure of an abstract mathematical system.
(2) Making inferences in the abstract system based on particular conditions of the given problem.
(3) Interpreting the conclusions drawn in the abstract system in terms of the original problem.

While one or more of these steps may occur tacitly in an actual application, by detailed analysis it should be possible to reconstruct the total application process by filling in the tacit steps.

If given broad construction, the term "application" covers a great deal of mathematical activity—any situation in which a general principle is used to make deductions in a particular setting. For example, to solve $x^2 + 3x - 2 = 0$ the quadratic formula can be applied to yield $x = -3 \pm \sqrt{17} / 2$. To prove that $5.7 + 3 = 3 + 5.7$ the commutative property of addition can be applied. Such broad interpretation of the term "application" leads to a distorted view of the actual activity of mathematics classes and mathematics in general. It intrudes on activity more naturally considered "developing" a mathematical system. In each case above, specific properties of the number systems are established by deduction from more general properties.

Mathematical application (AM) was coded only when properties of one abstract mathematical system were used to solve problems in another system. For example, a theorem of group theory might be used to deduce the fact that every integer has a unique additive inverse as follows:

1. The set of integers under addition form a commutative group.
2. In any group, each element has a unique inverse.
3. Therefore, each integer has a unique, additive inverse.

Using this restrictive definition, application will appear to be a much less common classroom activity than other studies have indicated. However, the definition chosen seems more appropriate in light of what mathematicians mean by the word "application."

The preceding explanations and examples give a general description of mathematical activity coding. More specific coding procedures and other examples are presented in the Appendix: "Instructions for Coding."

Logical Process

There is one important dimension of the coding process that has not yet been discussed. In the brief exchange

T: What is 7 times 5? (T/SOL/1/Wm/R2.21)
S: 35. (S/RES/1/Wm/R2.21)
T: 35 is right. (T/REA/1/Wm/R2.21)

the purpose of the first two utterances is clear from the coding of move and mathematical activity. The teacher asks for recall of a computational fact, and the student responds by stating the desired fact. But what is the function of the teacher's reaction? This must be described by another, non-mathematical, attribute: *evaluation.*

In the exchange

S: Why is (25 x 38) x 76 equal to (38 x 25) x 76? (S/SOL/2/Wm/R2.21)
T: Because the right side can be gotten from the left by use of the commutative property of multiplication. (T/RES/4/Wm/R2.21)

the teacher's response is directed to a specific computational fact, but it performs the function of *justifying* the fact by deduction.

In addition to developing mathematical ideas, each move in classroom discourse takes the form of what Smith has called a *logical operation.* According to Smith, moves in verbal discourse exhibit recurrent patterns which can be analyzed from a logical standpoint by comparing them to ideal models.[10] For example, logicians specify certain formal criteria that a good definition or deductive argument should meet. Even though moves in verbal communication seldom assume the ideal form called for by logic, they do exhibit reliably identifiable structure.

The studies of Smith and Bellack[11] have produced sets of logical operations found in classroom discourse. However, categories appropriate for analyzing logical operations in mathematics classes must reflect the more formal usage of the term "logical" in mathematics. The following procedures for coding logical process in discourse about mathematics are a modification of the approach developed in the Bellack study.

Analytic Process. "Analytic statements are statements about the proposed use of language."[12] These statements define or interpret words, symbols, or phrases by using combinations of words or symbols considered equivalent to the original.

A move is coded *defining* (Def) if and only if it gives or solicits "the set of properties or characteristics that an object (abstract or concrete) must have for the term to be applicable."[13] This restriction to what is commonly called *connotative definition* was necessary to make the analysis consistent with accepted mathematical use of the term "definition." If, in addition to giving defining characteristics, a speaker cites an example of the defined class (dennotative definition), the move is still coded Def.

However, a move is coded *interpreting* (Int) if it clarifies the meaning of either (1) a word or symbol by denoting an object (abstract or concrete) to which the word or symbol refers, or (2) a statement by rephrasing the statement in different but equivalent form.

The following excerpts from tapescripts illustrate the two kinds of analytic process:

T: Z will include all the strictly positive integers, all the strictly negative integers, and 0. (coded Def)

T: When you say, "Well, he's very conventional," what does that mean? He likes to go to conventions? (coded Int)

S: It means he's sort of very typical of all people or he is conservative. (coded Int)

T: Give me an example of an integer having an additive inverse. (coded Int)

Factual Process. In the Bellack study statements which "give information about the world based on one's experience with it,"[14] are called empirical statements. In mathematics facts are not usually verifiable by physical experimentation and observation. However, two empirical processes, *fact-stating* and *conditional-inferring,* do occur in the abstract world of mathematics.

A move was coded *fact-stating* (Fac) if it solicited or gave "an account, description or report of an event or state of affairs. To state a fact is to state what is, what was in the past, or what will be in the future."[15] For example:

T: What is $^{-}3 \times {}^{-}2$? (coded Fac)
S: That is really just the mapping n——> 2n. (coded Fac)

A move was coded *inferring* (Inf) if it solicited or stated a conclusion that followed from given conditions. For example, the brief exchange

T: If the product ab = 0, what must be true of a or b? (coded Inf)
S: One or both must be 0. (coded Inf)

is typical of many moves involving conditional–inferring.

Evaluative Process. Evaluative statements are statements which express or solicit judgment about the truth of a fact or the value or appropriateness of an idea or action. Moves which served such a purpose were coded *evaluation* (Val). For example:

S: I think what Joanne said was right. (coded Val)
T: Yes, addition is commutative in Z_5. (coded Val)

The second example illustrates a common problem in coding logical process. In one context, the teacher's remark might simply be a statement of fact in response to a student's question and thus should be coded Fac. However, in a different context it might be a reaction affirming the truth of another person's statement. Context is the determining factor in choosing the appropriate category. For instance, when a teacher repeated a student's statement (with or without adding some rating) the move was coded (Val) under the assumption that the very repetition represented some statement about the truth, falsity, or at least appropriateness of the preceding statement.

Justifying Process. Closely associated with factual and evaluative processes are the processes which support the truth of facts stated or judgments made. Warrants for factual statements or value judgments are commonly of three types: (1) logical deduction, (2) induction, or (3) authority or opinion.

A move was coded *deduction* (Ded) when it solicited or gave logical justification for a statement, inference, or evaluation. For example:

T: Why does $a + b = b + a$ for all a, b in Z? (T/SOL/2/Za/R2.22/Ded)
S: Because addition is commutative in the integers. (S/RES/2/Za/R2.22/Ded)

A move was coded *induction* (Ind) when it solicited or gave support for a statement or inference by citing similar situations or special cases implying a general justifying principle. For example:

T: How did you know that the product $^{-}1 \times 3$ should be defined to be $^{-}3$? (T/SOL/3/Zm/D2.21/Ind)
S: Well the products are all decreasing by 3(9, 6, 3, 0), so it must be $^{-}3$. (S/RES/3/Zm/D2.21/Ind)

A move was coded *opinion* (Opn) when it solicited or gave personal opinion or cited an authority (teacher or textbook, for example) as justification for a statement or inference. For example:

T: The book says only whole numbers can be used in ordered pairs. (coded Opn)
S: I don't care what the book says. I think we ought to be able to use negative numbers too. (coded Opn)

Opining moves were distinguished from evaluative moves by intent. Moves coded Opn served specifically to justify a point of view; moves coded Val merely expressed a point of view.

Typically, justifying exchanges contain two steps: (1) a solicitation of the justification, and (2) a response giving the justification. Usually the solicitor does not specify the type of justification sought. If context does not indicate the kind of justifying response desired, code the SOL as general justifying (Jus).

Cycles in Classroom Communication

Coding moves according to mathematical activity and logical process adds another important dimension to the analysis of verbal communication in mathematics classes. Among the interesting questions which such coding answers are the following:

1. What are the relative frequencies of occurrence of the three major classes of mathematical activity: developing, examining, and applying?
2. Do some classes place more emphasis than others on the formal aspects of mathematics such as logic and deduction?

3. Do mathematics class discussions involve substantially more logical deduction than discussions in social studies or literature classes?
4. What are the probabilities of transition from one kind of logical process to another; for example, from conditional inferring to deductive justification?

These and other similar questions are considered in Chapter III, which reports the results of data analysis. However, there are several important questions not answered by examination of moves alone. The classroom communication game is a sequential interaction in which each move is influenced by the preceding discussion and, in turn, influences the discussion that follows. If one studies only individual moves in isolation, the developmental structure of the discourse is left unexamined; the influence of the structure of mathematics on this development is unmeasured.

Although it is difficult to define units of analysis consisting of sequential sets of moves which can be reliably identified, the system of analyzing moves can be used for this purpose in a very reasonable and practical way. The first move in a given class discussion—when coded according to type of move, content, mathematical activity, and logical process—can be viewed as determining the state of the classroom communication system at the beginning of the class. The next move then represents the second state that the system enters. The difference between the code of move one and the code of move two is the change of state that has occurred. For instance, if the first move in a class meeting is:

T: Today we are going to take time to develop formally the definition of multiplication of integers. (T/STR/4/Zm/D2.1/Fac)

the teacher has structured the discussion by focusing attention on the topic—multiplication of integers—of the day's class. If the second and third moves are:

T: If x is any integer, what should the product 0·x equal? (T/SOL/2/Zm/D2.1/Inf)
S: 0. (S/RES/1/Zm/D2.1/Inf)

the content and mathematical activity remain constant, but the move type and the logical process change. Thus the state of the discussion moves from STR to SOL to RES and from Fac to Inf. If the next move is:

T: What property of (W,·) suggests that this should be the case? (T/SOL/3/Wm/R2.22/Ind) then the content and mathematical states have shifted to whole number multiplication and recall of properties of (W,·) respectively and the discussion might continue in this direction for several more moves.

The type of move and logical process commonly change from one move to the next. However, content and mathematical activity often stay constant for a set of moves developing a single idea. By dividing discourse into content and mathematical activity units, it becomes possible to analyze the pedagogical and logical structure of the way that mathematical ideas are developed.

DEFINITION: A *cycle* in classroom discourse is an uninterrupted sequence of moves for which one or more of the coding measures remains constant.

Using this basic definition, a content cycle is defined to be a cycle in which mathematical or non-mathematical content is the same; a mathematical activity cycle is one in which the mathematical activity is the same; a variety of other cycles can be defined by looking at the mathematical activity codes at different levels of specificity—an operation cycle, a definition of mapping cycle, and so on.[16]

The following excerpt from a transcript illustrates a mathematical activity cycle, in this case comparing systems:

T: The problem is to devise rules for multiplication in Z./ We have had several isomorphisms already. We have here W and Z. We set up several one-to-one correspondences./ What is the number in Z that corresponds to 8 in W, Nina?/

(T/STR/3/Zm/D2.1/Fac)
(T/STR/5/WZ/C1.0/Fac)
(T/SOL/3/WZ/C1.0/Fac)

S: Positive 8.

(S/RES/1/WZ/C1.0/Fac)

T: Positive 8./ Suppose we consider a problem in W, say 4 + 7. What would be its equivalent expression in Z, Mary?

(T/REA/1/WZ/C1.0/Val)
(T/SOL/4/WZ/C2.21/Fac)

S: Positive 4 plus positive 7.

(S/RES/1/WZ/C2.21/Fac)

T: What number in Z does this numeral name?

(T/SOL/2/WZ/C2.21/Fac)

S: Positive 11.

(S/RES/1/WZ/C2.21/Fac)

T: What number in W does this numeral name?

(T/SOL/2/WZ/C2.21/Fac)

S: 11.

(S/RES/1/WZ/C2.21/Fac)

T: 11 is right./ Now we want to use the similarity of (Z,+) and (W,+) to guide our definition of multiplication in Z.

(T/REA/1/WZ/C2.21/Val)
(T/STR/4/Zm/D2.1/Fac)

While illustrating the concept of mathematical activity cycle, the preceding excerpt also poses a problem. The eleven-move comparing-cycle is clearly an aside in the course of a discussion developing (Z, ·). However, the coding indicates only that a cycle developing (Z, ·) was followed by a cycle comparing (Z, +) and (W, +) which was then followed by a cycle developing (Z, ·). To avoid such deceptive coding, content and mathematical activity are coded on two levels. The first, or general, level indicates the basic substantive topic and mathematical activity under consideration in the discussion taken as a whole. The second, or specific, level indicates the content and mathematical activity involved in some distinct sub–cycle of the discussion. It is conceivable that an intricate hierarchy of coding levels could result if this differentiation between general and specific content and activity is extended to ever finer distinctions. However, two levels seemed adequate to describe the situations commonly met in actual transcripts.

Further illustrations and specific coding procedures for recognizing sub–cycles are presented in the Appendix: "Instructions for Coding." A detailed analysis of the structure of content and mathematical activity cycles appears in Chapter III.

Special Coding Problems

The move, mathematical activity, and logical process definitions were decided upon only after examination of large portions of the protocols. However, several coding problems persisted: recurrent situations exhibited ambiguity of meaning that could not be reliably coded.

Two coding problems centered around the structuring move. The first, anticipated by Bellack's experience, involved identifying the end of a reacting move and the beginning of a new structuring move. Detailed discussion of each move appears in the special instructions for coders. In general,

> If doubt persisted with regard to differentiating structuring and reacting moves, the general rule was to code that pedagogical category which moved the discourse forward, thus coding structuring rather than reacting.[17]

The second structuring move problem is illustrated by the move:

T: Let's assume x is 3 and y is 2. Then what is $x + y$?

Although the first complete sentence in this utterance is a statement and thus codable as STR, it is really part of the question that follows, a SOL. Short statements which merely stated hypotheses for a following dependent SOL were coded as part of that SOL. Longer introductory statements (two or more complete sentences) were coded STR.

Difficult decisions appear in coding mathematical activity when a set of moves may or may not constitute a sub–cycle of a general activity. For instance, proof of a new result often involves recall of previously established facts. Do the recall moves constitute a distinct sub–activity, or are they to be coded as part of the general activity which is developing? In general, a set of moves was considered a sub–cycle when, taken as a unit, these moves were meaningful and independent of the main context (yet contributed to that discussion), or when the main discussion was meaningful with the sub–cycle removed. Under this interpretation, recall moves which were an integral part of the proof of some new fact were not coded as a sub–cycle.

The main logical process coding problems are of two types. The first arises from difficulty in distinguishing between Fac and Val such as:

T: Is + commutative in W?

where the SOL might be considered as seeking a statement of fact or merely a judgment of truth or falsity (thus Val). In general, Val is coded only if the move solicits or gives rating of a previous statement. The second arises from difficulty in distinguishing between Fac and Inf. Frequently SOL moves that have the appearance of hypothesizing are actually simply requests for statements of fact. For example:

T/SOL: If x is 3 and y is 2, then $x + y$ is?

This teacher's SOL can in one sense be considered as calling for an inference. On the other hand, it can be considered as purely a call for statement of fact: $3 + 2 = 5$. In general, regardless of the occurrence of an "if p then —" form, Inf was coded only when several appropriate conclusion were possible, or when the given or solicited conclusion was a fact not previously known to the class.

In all cases where it was impossible to distinguish between two or more sub–categories of a process, the move was coded as simply *analytic process* (Aly), *factual process* (Fap), or *justifying process* (Jus). When more than one logical process occurred in a single move, the code was chosen according to an order of precedence which is given in the instructions for coders.

These represent the typical and most common coding problems encountered. They were not, however, the only difficulties faced in describing classroom discourse about mathematics; each coder had to examine the utterances very carefully before inferring the appropriate move, content, activity, and logical process code.

RELIABILITY TEST

To determine whether the system described in this chapter could be used reliably, a sample of previously unexamined protocols was coded by independent teams and the coding was compared. The 20 protocols constitute approximately 400 pages of data. In determining a reasonable portion of this data to be analyzed in the reliability test, two factors were considered.

Because of the highly technical nature of the discussion, coders were required to be completely familiar with the mathematics being discussed, as well as with the relation of this material to previous topics in the SSMCIS course. Thus, the more segments chosen, the greater the problem for coders to assimilate the context, within which each segment of discourse took place. On the other hand, if only a few segments had been chosen, it was possible that many mathematical activity and content categories would not have been represented in the sample.

As a compromise between these potential difficulties, six segments of five pages each were chosen at random from the protocols. Then these segments were independently coded by two teams of coders trained in the use of the system of analysis. The coding done by the two teams was then compared to determine the per cent of agreement on move, content, mathematical activity, and logical process coding.

The coding teams consisted of doctoral students in mathematical education (two on each team) who had no prior experience with the system of analysis, with the SSMCIS seventh–grade course, or with other similar schemes for analyzing classroom verbal interaction. As anticipated, the coders found it difficult to analyze mathematical activity in segments selected at random from the transcripts of 20 different class sessions. The logically ordered nature of mathematics made this difficulty especially crucial because the purpose of any given activity had to be described in the context of a full year of activity.

The per cent of agreement in coding was highest in the pedagogical move category, 85% of all moves being coded the same by the two teams. Counting only disagreements about major logical process category (analytic, factual, evaluative, or justifying), the teams agreed on the logical purpose of 82% of all moves. If disagreements about intra–category refinements are to be counted (e.g., definition versus interpretation), agreement falls to 74% of all moves.

The coding teams agreed on content codes for 93% of all moves, all disagreement being in a single segment of moves in one protocol. The percent of agreement on mathematical activity codes was 79%, but again, the disagreements that occured were confined to only a few segments because mathematical activity tended to remain constant for long chains of moves.

The high per cents of agreement are encouraging, but the experience indicates that the system of analysis cannot be used casually, or by persons with a limited grasp of the structure and content of mathematics.

3

VERBAL COMMUNICATION IN SELECTED MATHEMATICS CLASSES

Each aspect of the system of analysis was designed to measure an important dimension of classroom verbal behavior. Thus, when used to examine transcripts of the 20 tape-recorded class sessions, the coding produced a profile of verbal activity in those classes. Within this profile, various patterns of source, duration, pedagogical purpose, content, mathematical activity, and logical process emerged: in individual class sesessions, in classes of particular teachers, and in the protocols taken as a whole.

OVERVIEW OF VERBAL ACTIVITY

The 20 recorded sessions (four sessions each of classes taught by five teachers) produced a total of 9,506 moves and 20,381 lines. These represent averages of 475 moves per session and 1,019 lines per session. However, examination of the activity in individual classes shows considerable deviation from these averages. For instance, the per session averages for class five were 283 moves and 731 lines, whereas class four averaged 568 moves and 1,142 lines per session.

This deviation can be explained partly by the fact that class five met for an average of 39 minutes per session while class four averaged 48 minutes per session. But even adjusting for this difference, class five showed a markedly lower rate of activity than class four.

Table I gives the average number of moves per session, lines per session, lines per move, moves per minute, and lines per minute for each class observed and for all classes combined ($\bar{x}$).

TABLE I

Average number of moves and lines, moves per minute, lines per minute, and lines per move for each class and for all classes combined

	Class					
	1	2	3	4	5	$\bar{X}$
moves	500	551	475	568	283	475
lines	971	1140	1110	1142	732	1019
lines/move	1.9	2.1	2.3	2.0	2.6	2.1
moves/minute	11.1	11.5	8.6	11.9	7.3	10.1
lines/minute	21.6	23.8	20.2	24.0	18.9	21.7

In addition to variations from class to class, the rates of lines per move, moves per minute, and lines per minute fluctuated from session to session in a single class. However, as Table II shows, there was little apparent pattern to this fluctuation.

SOURCES OF VERBAL ACTIVITY

The patterns of rate and production of verbal activity are more interesting when refined to indicate distribution according to source. In each class and for each session of each class, the teacher spoke more than all his students combined. Overall, the ratio of teacher to student talk was about 3 : 2 in terms of moves and 5 : 2 in terms of lines. The slightly higher ratio for lines is explained by the fact that the average teacher move was 2.6 lines long and the average student move 1.5 lines long. Not only did teachers speak more often than students, but each time they spoke they did so for a longer time.

As was the case for general productivity of verbal communication, teacher–student ratios in individual classes varied from the averages for all classes combined. This variation was most pronounced in terms of lines spoken. In class five the teacher spoke an average of 82% of all lines to only 18% for students, or a ratio of 9 : 2. In contrast, the teacher in class three spoke an average of

TABLE II

Average number of lines per move, moves per minute, and lines per minute for each session of each class and for all classes combined

Class		Session I	II	III	IV
1.	lines/move	2.3	2.0	1.8	1.8
	moves/minute	10.2	9.6	13.2	11.4
	lines/minute	23.3	19.1	23.6	20.2
2.	lines/move	2.1	2.0	2.2	1.9
	moves/minute	9.7	11.7	10.3	14.2
	lines/minute	19.9	24.0	23.6	27.4
3.	lines/move	2.7	2.1	2.5	2.0
	moves/minute	6.9	8.9	8.8	9.8
	lines/minute	18.8	20.1	22.1	19.8
4.	lines/move	1.9	2.1	2.2	1.8
	moves/minute	13.3	11.7	11.7	11.0
	lines/minute	25.4	24.9	25.8	19.8
5.	lines/move	2.6	2.4	2.6	2.9
	moves/minute	6.8	9.0	6.6	6.7
	lines/minute	17.8	21.2	17.5	19.2
$\overline{X}$.	lines/move	2.2	2.1	2.2	2.0
	moves/minute	9.4	10.2	10.2	10.7
	lines/minute	21.0	21.9	22.5	21.3

65% of all lines to 35% for students, or a ratio of approximately 2 : 1. This difference is also indicated by a comparison of the lines per move ratios for the two classes.

	Class 3		*Class 5*	
	T	*S*	*T*	*S*
lines/move	2.7	1.9	3.3	1.3

Per cent of moves and lines for teachers and students in each class and in all classes combined is shown in Table III.

TABLE III

% of lines and moves for teachers and students in each class and in all classes combined

Class	*Teacher*	*Student*
1. Moves	62.5	37.5
Lines	76.1	23.9
2. Moves	61.4	38.6
Lines	73.2	26.8
3. Moves	56.0	44.0
Lines	65.1	34.9
4. Moves	55.5	44.5
Lines	67.9	32.1
5. Moves	65.8	34.2
Lines	82.3	17.7
$\overline{X}$ Moves	60.2	39.8
Lines	72.9	27.1

Variation between classes in the teacher–student activity ratio was not accompanied by any pattern of differences among the several sessions of individual classes. For example, in class one the teacher spoke an average of 63% of all moves. In individual sessions of this class, the teacher spoke 65%, 62%, 61%, and 63% of all moves. Table IV gives the overall picture; the irregularities that occur form no apparent pattern. The teacher dominated verbal activity in all classes; this domination was consistent from class to class and from session to session within a single class.

TABLE IV

% of lines and moves for teachers and students in each class, each session

	Teacher					*Students*				
	Session					*Session*				
Class	I	II	III	IV	$\overline{X}$	I	II	III	IV	$\overline{X}$
1. Moves	64.5	61.6	60.9	62.9	62.5	35.5	38.4	39.1	37.1	37.5
Lines	78.9	72.1	75.2	77.6	76.0	21.1	27.9	24.8	22.4	24.0
2. Moves	61.8	61.0	62.8	60.7	61.6	38.2	39.0	37.2	39.3	38.4
Lines	75.3	70.9	74.2	73.1	73.4	24.7	29.1	25.8	26.9	26.6
3. Moves	58.1	50.2	57.3	58.3	56.0	41.9	49.8	42.7	41.7	44.0
Lines	76.0	50.3	68.2	65.9	65.1	24.0	49.7	31.8	34.1	34.9
4. Moves	53.9	56.8	55.8	55.3	55.5	46.1	43.2	44.2	44.7	44.5
Lines	68.3	67.4	70.5	65.3	67.9	31.7	32.6	29.5	34.7	32.1
5. Moves	68.3	65.0	64.5	65.5	65.8	31.7	35.0	35.5	34.5	34.2
Lines	85.8	80.4	74.3	86.6	82.3	14.2	19.6	25.7	13.4	17.7
$\overline{X}$. Moves	61.3	58.9	60.3	60.5	60.3	38.7	41.1	39.7	39.5	39.7
Lines	76.9	68.2	72.5	73.7	72.9	23.1	31.8	27.5	26.3	27.1

The patterns of rate, productivity, and distribution according to source of verbal activity in the observed junior high school mathematics classes, are in some ways similar and in other ways quite different from those found by Bellack in senior high school social studies classes. First, the social studies classes averaged only about 50% as many moves per session, but 50% more lines, and nearly three times as many lines per move as the mathematics classes.[1] Discussion in eleventh- and twelfth-grade social studies classes apparently consists of fewer but longer utterances than discussion in seventh- and eighth-grade mathematics classes.

TABLE V

Comparison of rates of verbal activity in mathematics and social studies classes

	Junior High School Mathematics	*Senior High Social Studies*
moves/session	475	250
lines/session	1,019	1,418*
lines/move	2.2	5.7

*Adjusted to account for different line unit used.

Whether it is the difference in age of students or the difference in substantive content under discussion that is primarily responsible for these differences can only be conjectured. Undoubtedly older students have greater verbal facility, but it is not unreasonable to expect that the more subjective nature of questions in social studies would encourage longer individual utterances than those needed to talk about mathematics.

With respect to distribution of verbal activity between teachers and students, the social studies and mathematics classes are much more similar. In both, the teacher–student ratio is approximately 3 : 2 in terms of moves and 3 : 1 in terms of lines. Thus, while the rate and production of verbal activity may be affected by subject matter or age of students, the relative contributions of teachers and students appear to be invariant.

PEDAGOGICAL PURPOSES OF VERBAL ACTIVITY

Each move in classroom discourse serves one of four pedagogical purposes: structuring, soliciting, responding, or reacting. In the observed mathematics classes, approximately 5% of all moves were structural, 32% solicited information or action, 32% were responses to solicitations, and 31% were reactions to a prior move or moves. Table VI shows that the distribution according to lines was somewhat different; structuring moves tended to be more than twice as long as either soliciting or reacting moves and more than three times as long as responding moves.

TABLE VI

% of moves and lines devoted to each type of pedagogical move and ratio of lines per move

	% of Moves	*% of Lines*	*Lines/Move*
Structuring	5.2	12.2	5.2
Soliciting	32.1	33.3	2.3
Responding	31.7	21.5	1.5
Reacting	31.0	33.0	2.3

Analysis of the distribution of moves according to pedagogical purpose and source gives unmistakable evidence of a dichotomy between teacher and student roles in the classroom. Whether measured in terms of moves or lines, teachers did most of the structuring, soliciting, and reacting. The student was expected mainly to respond to teacher solicitations and only occasionally structure, react, or ask a question of his own. Table VII shows the comparisons.

These patterns in frequency and source of the pedagogical moves suggest that classroom discourse is a sequence of episodes, each consisting of a teacher solicitation, followed by a student response and a teacher reaction. Only 5% of these episodes were initiated by teacher structuring, while an even smaller per cent included some sort of student reaction following one by the teacher. Predominance of this pattern is confirmed by Table VIII, which gives the most common move types that followed structuring, soliciting, responding, and reacting moves, respectively.

TABLE VII

Distribution of move types according to source

	% of Moves by Teacher	*% of Lines by Teacher*	*% of Moves by Student*	*% of Lines by Student*
Structuring	80.3	86.0	19.7	14.0
Soliciting	92.6	93.4	7.4	6.6
Responding	4.4	9.9	95.6	90.1
Reacting	82.9	88.2	17.1	11.8

TABLE VIII

Most common initial following move transitions

Initial move type	*Following move type*	*Relative frequency of following move type given initial move type*
STR	SOL	.66
SOL	RES	.90
RES	REA	.66
	SOL	.28
REA	SOL	.58
	REA	.24

Patterns in Individual Classes

As Tables IX - XII show, the trend of teacher domination of the structuring, soliciting, and reacting moves was maintained consistently in each class and for each session of each class.

The single exception to this rule was session II of class three in which students made 71% of the structuring moves (27 of 38). Even in this session, however, the teacher maintained his dominant role in soliciting and reacting.

When one recalls that these teacher–student ratios do not represent a dialogue between the teacher and a single student, but an interaction between a single teacher and 20 to 30 students, the role of each individual student in shaping classroom discourse appears particularly small.

TABLE IX

% of structuring moves and lines by teachers and students in each session of each class

Class	Teacher					Students				
	I	II	III	IV	$\overline{X}$	I	II	III	IV	$\overline{X}$
1. Moves	88.8	100.0	92.0	95.5	94.1	11.2	0.0	8.0	4.5	5.9
Lines	98.1	100.0	95.6	99.2	98.2	1.9	0.0	4.4	0.8	1.8
2. Moves	86.7	80.0	81.6	87.5	83.9	13.3	20.0	18.4	12.5	16.1
Lines	95.8	88.9	90.2	91.0	91.5	4.2	11.1	9.8	9.0	8.5
3. Moves	78.9	28.9	51.6	60.9	55.1	21.1	71.1	48.4	39.1	44.9
Lines	93.6	26.7	55.6	64.6	60.1	6.4	73.3	44.4	35.4	39.9
4. Moves	64.9	73.3	70.0	74.3	70.6	35.1	26.7	30.0	25.7	29.4
Lines	83.5	82.9	76.5	83.3	81.5	16.5	17.1	23.5	16.7	18.5
5. Moves	100.0	100.0	90.9	100.0	97.7	0.0	0.0	9.1	0.0	2.3
Lines	100.0	100.0	94.7	100.0	98.7	0.0	0.0	5.3	0.0	1.3

TABLE X

% of soliciting moves and lines by teachers and students in each session of each class

Class		Teacher					Students				
		I	II	III	IV	$\overline{X}$	I	II	III	IV	$\overline{X}$
1.	Moves	96.4	95.1	95.3	97.1	96.0	3.6	4.9	4.7	2.9	4.0
	Lines	98.4	90.3	96.3	99.2	96.0	1.6	9.7	3.7	0.8	4.0
2.	Moves	96.5	96.5	96.1	96.4	96.4	3.5	3.5	3.9	3.6	3.6
	Lines	97.4	97.6	97.2	97.0	97.3	2.6	2.4	2.8	3.0	2.7
3.	Moves	83.8	96.7	85.4	91.9	89.2	16.2	3.3	14.6	8.1	10.8
	Lines	86.3	88.7	87.6	93.5	89.0	13.7	11.3	12.4	6.5	11.0
4.	Moves	85.0	85.8	82.1	76.4	82.3	15.0	14.2	17.9	23.6	17.7
	Lines	89.2	83.5	86.5	82.2	85.3	10.8	16.5	13.5	17.8	14.7
5.	Moves	98.8	100.0	98.9	98.6	99.1	1.2	0.0	1.1	1.4	0.9
	Lines	99.2	100.0	99.6	99.5	99.6	0.8	0.0	0.4	0.5	0.4

TABLE XI

% of responding moves and lines by teachers and students in each session of each class

Class	Teacher					Students				
	I	II	III	IV	$\bar{X}$	I	II	III	IV	$\bar{X}$
1. Moves	2.1	2.7	2.2	1.8	2.2	97.9	97.3	97.8	98.2	97.8
Lines	2.9	9.1	2.5	2.8	4.3	97.1	90.9	97.5	97.2	95.7
2. Moves	2.0	3.1	2.0	1.5	2.2	98.0	96.9	98.0	98.5	97.8
Lines	4.9	3.2	5.0	12.2	6.3	95.1	96.8	95.0	87.8	93.7
3. Moves	9.2	3.0	10.9	4.4	6.9	90.8	97.0	89.1	95.6	93.1
Lines	32.5	2.3	12.7	3.9	12.9	67.5	97.7	87.3	96.1	87.1
4. Moves	8.3	6.9	10.8	12.6	9.7	91.7	93.1	89.2	87.4	90.3
Lines	15.6	23.4	34.7	21.3	23.8	84.4	76.6	65.3	78.7	76.2
5. Moves	2.4	0.0	0.0	1.3	0.9	97.6	100.0	100.0	98.7	99.1
Lines	6.7	0.0	0.0	1.1	2.0	93.3	100.0	100.0	98.9	98.0

TABLE XII

% of reacting moves and lines by teachers and students in each session of each class

	Teacher					Students				
Class	I	II	III	IV	$\bar{X}$	I	II	III	IV	$\bar{X}$
1. Moves	91.2	89.3	79.3	86.9	86.7	8.8	10.7	20.7	13.1	13.3
Lines	95.7	95.1	86.1	91.4	92.1	4.3	4.9	13.9	8.6	7.9
2. Moves	86.9	93.6	84.9	85.0	90.1	13.1	6.4	15.1	15.0	9.9
Lines	91.9	94.4	89.8	90.8	91.7	8.1	5.6	10.2	9.2	8.3
3. Moves	72.0	60.8	70.3	71.6	68.7	28.0	39.2	29.7	28.4	31.3
Lines	83.5	66.0	85.9	82.5	79.5	16.5	34.0	14.1	17.5	20.5
4. Moves	68.8	77.8	71.8	71.9	72.6	31.2	22.2	28.2	28.1	27.4
Lines	76.4	84.2	79.2	79.3	79.8	23.6	15.8	20.8	20.7	20.2
5. Moves	97.4	96.0	94.8	96.7	96.2	2.6	4.0	5.2	3.3	3.8
Lines	98.0	98.4	96.7	98.8	98.0	2.0	1.6	3.3	1.2	2.0

Despite the dominance of teacher leadership in all classes, some interesting differences can be detected between individual classes with respect to teacher–student ratios for each move type. For instance, in class five, students did almost no structuring. On the other hand, in class three, the ratio of teacher to student structuring was only 5 : 4 in terms of moves and 3 : 2 in terms of lines. Similar contrasts appeared in the distribution of soliciting, responding, and reacting moves and lines.

TABLE XIII

Comparison of classes three and five according to source of pedagogical move types

	Class 3		*Class 5*	
	Teacher	*Students*	*Teacher*	*Students*
Structuring				
moves	55.1	44.9	97.7	2.3
lines	60.1	39.9	98.7	1.3
Soliciting				
moves	89.2	10.8	99.1	0.9
lines	89.0	11.0	99.6	0.4
Responding				
moves	6.9	93.1	0.9	99.1
lines	12.9	87.1	2.0	98.0
Reacting				
moves	68.7	31.3	96.2	3.8
lines	79.5	20.5	98.0	2.0

Although the per cent difference between distribution of the move types in the two classes is most striking as regards structuring moves, it must be kept in mind that structuring represents only about 5% of all moves and 12% of all lines in classroom discourse. Thus, it appears that when students were allowed to break out of a monotone responding role, it was mostly by reacting, and only occasionally by structuring, rather than by reversing the dominant T/SOL - S/RES pattern.

COMPARISON OF PEDAGOGICAL PATTERNS IN MATHEMATICS AND SOCIAL STUDIES CLASSES

As in the Bellack study, the concept of pedagogical move served as a useful unit of analysis in the study of patterns of verbal communication in the mathematics classes. The classes participating in the present study were all pilot classes of the SSMCIS working on a chapter entitled "Multiplication of Integers." Although all teachers had attended a summer course to prepare them to teach the new SSMCIS material, most of this instruction was devoted to mathematical rather than pedagogical matters, so it is safe to assume that they had acquired little uniformity of teaching style as a result of this instrúction. Furthermore, the teachers were given no special instructions prior to the tape–recording except to "be themselves."

It was found that there was an evident pattern of pedagogical moves which held for each class and for each session of each class. Approximately 4% of all moves (11% of the lines) were devoted to teacher structuring; 30% of all moves (31% of the lines) to teacher soliciting; 31% of all moves (19% of the lines) to student responding; and 25% of all moves (29% of the lines) to teacher reacting. The balance of class time was taken up by occasional student structuring, soliciting, and reacting.

These findings compare favorably with the patterns discovered by the Bellack study of senior high school social studies classes. In those classes, teachers also did a majority of the structuring, soliciting, and reacting, leaving the responding role to their students. The teacher–student ratios in terms of moves were 12 : 1 for structuring, 13 : 2 for soliciting, 1 : 7 for responding, and 4 : 1 for reacting.[2] This resulted in approximately 5% of all moves (15% of the lines) being devoted to teacher structuring, 29% of all moves (20% of the lines) to teacher soliciting, 25% of all moves (16% of the lines) to student responding, and 24% of all moves (32% of the lines) to teacher reacting.

The striking similarity in relative frequency and sequencing of the various move types in senior high school social studies classes

TABLE XIV

Comparison of patterns of pedagogical moves in junior high school mathematics classes and senior high school social studies classes. % of all moves and all lines devoted to each move type by teachers and students

	Junior High School Mathematics		*Senior High School Social Studies*	
	Teacher	*Students*	*Teacher*	*Students*
Structuring				
Moves	4.1	1.1	4.8	0.4
Lines	10.6	1.6	14.5	3.0
Soliciting				
Moves	29.6	2.5	28.8	4.4
Lines	31.2	2.0	20.3	2.5
Responding				
Moves	1.2	30.5	3.5	25.0
Lines	2.3	19.3	5.0	15.6
Reacting				
Moves	25.1	5.9	24.3	5.7
Lines	28.9	4.1	32.0	5.1

and junior high school mathematics classes would, if supported by other research, be an interesting commentary on the way teachers naturally tend to run their classes.

CONTENT AND MATHEMATICAL PURPOSE OF VERBAL ACTIVITY

Mathematics has long been considered one of the most highly structured of all disciplines. The logical interdependence of content within specific branches of mathematics has been known at least since Euclid cast geometry in postulational form, and modern developments have exposed the unity of concepts and methods underlying all traditional branches of mathematics. It seems reasonable to expect that the logical organization of mathematics imposes restrictions on any classroom discussion about mathematics. With few exceptions, the analysis of content development and mathematical activity (and later the structure of content cycles) in the observed classes supports this hypothesis.

Content Emphasis

The number systems of arithmetic can be developed in a variety of mathematically equivalent ways. However, once a path has been chosen (as embodied in a textbook) and followed part way, variation is restricted by the order of logical dependence among the properties of the systems.

Students in the observed classes had earlier studied addition and the integers, and had been shown that the set of non-negative integers is isomcrphic to the whole numbers under the operation of addition. The objective of the chapter "Multiplication of Integers" in the SSMCIS is to define multiplication of integers in a way that extends the isomorphism of $(W,+)$ and $(Z \geq o,+)$ to $(W,+,\cdot)$ and $(Z \geq o,+,\cdot)$. Sections 6.1 and 6.2 are a review of the basic properties of $(W,\cdot)$; sections 6.3 - 6.7 develop the definition and properties of $(Z,\cdot)$; sections 6.8 - 6.10 develop another isomorphism, this between the integral dilations of a line under the operation of composition and $(Z,\cdot)$; and sections 6.11 - 6.12 prove some theorems about $(Z,\cdot)$.

None of the classes could complete all 15 sections of the chapter in the four tape–recorded sessions. In fact, the slowest class got only as far as section 6.7 and the fastest to part of 6.12. However, the data showing content emphasis in successive sessions for all classes combined and for each class separately fit in the sequence embodied in the textbook very closely.

In session I, more than 50% of all lines and moves were devoted to consideration of whole number multiplication. By session II this had declined to 20%, and in session IV no reference was made to $(W,\cdot)$. On the other hand, emphasis on multiplication of integers grew from 11.8% of the moves (12% of the lines) in session I to 40.6% of the moves (40.6% of the lines) in session IV. As would be expected, dilations entered the discussion primarily in sessions III and IV. Thus, in general, the classes followed the sequence of topics in the textbook chapter.

Teachers and students also kept remarkably close to the subject of the chapter in all observed classes. Aside from discussion of classroom procedural matters (such as assignments) which took up about 5% of all moves and 6% of all lines, there was almost no discussion of unrelated substantive topics. The 14.2% of the moves (13% of the lines) in session IV that were coded Relevant Content, can be attributed almost entirely to discussion of one particular problem in section 6.2 which did not directly relate to the goals of the chapter.

The differences in content emphasis that occured from class to class were primarily a consequence of differences in teaching pace. One teacher (class three), whose students had been exposed to multiplication of integers in an earier course, moved quickly through the definitions, omitting all reference to the isomorphism between $(Z_{\geq 0},+)$ and $(W, +)$. Another (class five) moved more slowly, reviewing work in the earlier chapter on integers as he went. Despite these differences, the order of content emphasis was essentially the same in each class.

The discovery that all classes followed essentially the same path through the subject matter of the unit and spent a negligible amount of time discussing unrelated substantive topics, is in direct contrast to the patterns of content emphasis Bellack found in social studies classes. He observed that classes "showed marked differences in the substantive material covered in the class sessions." In fact, although teachers were given the same unit to teach during the tape–recorded sessions, "It almost seems as if the teachers were teaching a different unit." Those social studies teachers also devoted more class time to matters not directly concerned with the subject of the unit being studied. In one class, this proportion was almost 25% of the discussion.[4] This contrast between social studies and mathematics classes is yet further indication that the structure of mathematics as a discipline forces some structure on the pattern of discussion about mathematics.

Mathematical Activity

As was the case with patterns of content emphasis, the pattern of mathematical activity was roughly that which would be predicted by analysis of the textbook chapter being studied in the five mathematics classes. Text section 6.1 is designed to recall the properties of whole number multiplication and the isomorphism between $(W, +)$ and $(Z_{\geq 0},+)$. As predicted, in session I 46% of all moves and 43% of all lines were devoted to *recall* and *comparison.* In sessions II, III, and IV, the amount of time spent on these activities declined sharply. Emphasis on *developing* (definition and properties of $(Z,\cdot)$), already prominent in session I, doubled in session II. *Illustration* and *examination,* which can occur naturally either as part of recall or immediately following development of new ideas, were

TABLE XV

% of all moves and lines devoted to major* content topics in all classes combined by session

Content	Session I	II	III	IV
1. Whole Numbers (multiplication)				
moves	52.8	20.4	1.4	0
lines	52.6	23.8	2.4	0
2. Whole Numbers (general reference)				
moves	7.4	15.6	16.4	0
lines	6.8	16.6	13.8	0
3. Two Systems (compared)				
moves	7.0	0	0	1.8
lines	7.2	0	0	1.2
4. Relevant Content				
moves				
lines	0	14.2	5.3	0
	0	13.0	4.8	0
5. Integers (multiplication)				
moves	11.8	23.0	35.2	40.6
lines	12.0	23.6	36.0	40.6
6. Integers (general reference)				
moves	0	3.6	8.2	26.6
lines	0	3.2	9.2	26.2
7. Dilations				
moves	2.4	0	7.2	10.6
lines	3.4	0	6.4	11.4

* Remaining content categories represented by small percents.

distributed fairly evenly throughout the four sessions. As predicted in Chapter II of this study, *application* of mathematics was not a common activity in the observed classes.

The per cent of moves and lines devoted to each of the seven mathematical activities was roughly the same in all five classes—with two noticeable exceptions. Class three, which skipped review of (W, +), spent the most time developing new ideas (67% of the moves and lines) and the least time illustrating them (1% of both). Class four did the most illustrating (15% of the moves, 14% of the lines) and the least developing (46% of the moves, 47% of the lines).

In the present study, no attempt was made to measure the effect of such variation on student achievement. However, the differences suggest an interesting question: Do those teachers who emphasize illustration of new concepts get better student achievement than those who minimize illustration? The analysis of content and mathematical activity cycles presented later, shows that it is the teacher who is overwhelmingly responsible for choice of mathematical activity in the classroom who gets better results. Thus, significant findings can be translated directly into revised classroom procedure by altering teacher behavior.

Of far more interest than simple measures of relative frequency is the way in which different mathematical activities are put together in studying a single mathematical topic. Is development commonly preceded by recall? Does illustration lead to examination? Or, more specifically, does development commonly proceed from computation to general properties, or is the process reversed? These questions are considered in the discussion of content and mathematical activity cycles.

CONTENT AND MATHEMATICAL ACTIVITY CYCLES

The codes for source, pedagogical purpose, duration, and logical process commonly change from move to move. But the mathematical codes often continue unchanged for a sequence of related moves, thus forming a mathematical code cycle.[5]

In the observed mathematics classes, the mathematical code changed in some way an average of 23 times per session (disregarding interruptions by procedural cycles which occurred an average of nine times per session and lasted an average of eight lines each). The average mathematical code cycle was 42 lines in duration, but the median cycle length was actually less than 25 lines. As Table XVIII shows, there was considerable variation of these averages

TABLE XVI

% of moves and lines devoted to each type of mathematical activity in all classes combined by session

	Session				
Mathematical Activity	I	II	III	IV	$\bar{X}$
1. Developing					
moves	30	66	68	68	58.0
lines	33	65	65	67	57.5
2. Recall					
moves	39	15	9	6	17.3
lines	35	13	8	8	16.0
3. Illustration					
moves	13	2	10	12	9.3
lines	11	2	11	10	8.5
4. Comparison					
moves	7	0	1	2	2.5
lines	8	0	1	1	2.5
5. Examination					
moves	5	11	7	7	7.5
lines	5	13	8	9	8.8
6. Application (math.)					
moves	0	0	0	0	0
lines	0	0	0	0	0
7. Application (non-math.)					
moves	0	0	2	0	.5
lines	0	0	2	0	.5
8. Procedural					
moves	6	6	3	5	5.0
lines	8	7	5	5	6.3

TABLE XVII

% of moves and lines devoted to each mathematical activity in each class

Mathematical Activity	Class 1	2	3	4	5	$\overline{X}$
1. Developing						
moves	61	53	67	46	61	57.6
lines	61	54	67	47	59	57.6
2. Recall						
moves	13	20	16	19	19	17.4
lines	13	18	14	16	21	16.4
3. Illustration						
moves	6	12	1	15	13	9.4
lines	5	12	1	14	10	8.4
4. Comparison						
moves	2	1	1	6	3	2.6
lines	2	1	1	6	2	2.4
5. Examination						
moves	9	12	8	7	2	7.6
lines	9	12	11	9	2	8.6
6. Application (math.)						
moves	0	0	0	0	0	0
lines	0	0	0	0	0	0
7. Application (non-math.)						
moves	3	0	0	0	0	0.6
lines	2	0	0	0	0	0.4
8. Procedural						
moves	6	2	7	7	2	4.8
lines	8	3	6	8	6	6.6

from class to class and from session to session in a single class; but no pattern to this variation was apparent.

TABLE XVIII

Mathematical code cycles per session and lines per cycle in each session of each class and in all classes combined

	Session				
Class	I	II	III	IV	$\bar{X}$
1. Cycles	18	30	17	17	20
Lines/Cycle	51	26	60	48	43
2. Cycles	23	34	17	22	24
Lines/Cycle	38	32	65	57	48
3. Cycles	16	19	21	21	19
Lines/Cycle	59	54	54	49	54
4. Cycles	20	28	25	24	24
Lines/Cycle	58	37	48	33	44
5. Cycles	30	35	18	21	26
Lines/Cycle	21	21	37	32	28
$\bar{X}$. Cycles	21	29	20	21	23
Lines/Cycle	44	34	53	42	43

Earlier analysis of pedagogical move types according to source showed clearly that teachers play the dominant role in directing classroom discourse. This finding is confirmed by the distribution of moves which initiate or close mathematical code cycles. Teachers initiated almost 90% of all cycles, either by structuring (23%) or by soliciting (61%), and closed more than 80% of all cycles, primarily by reacting (63%). The only indication of student participation in directing the course of classroom discussion was occasional occurrence of a student response closing a cycle. However, when a student closed a cycle, this usually meant only that the teacher failed to react to the student's response and instead went on to a new topic, thus initiating a different mathematical code cycle. As can be seen in Table XIX, the pattern of teacher domination showed little variation from class to class or from session to session within each class.

The distribution of mathematical code cycles according to content and mathematical activity was discussed earlier. The important remaining question is whether or not the changes in mathematical code occur in any significant or predictable pattern.

The most informative way to approach this question is to study sequences of mathematical code cycles which have the same content sub–code; that is, to examine the internal structure of mathematical content cycles. The pattern of changes in mathematical activity during discussion of a single substantive topic provides further evidence of the way mathematical structure influences the structure of discourse about mathematics. For example, the following content cycle occured in class four:

Content	*Activity*	*Focus*	
D	D	3.10;	38 moves, 67 lines
D	D	3.21;	9 moves, 11 lines
D	D	3.22;	7 moves, 12 lines
D	I	3.22;	10 moves, 16 lines

This content cycle began with definition of a dilation mapping (3.10), proceeded to computing images (3.21), a general property of the mapping (3.22), and finally to illustrating (I) this property. To a certain extent, this pattern of change could be predicted by logical necessity; it is obviously necessary to define the mapping before computing images. However, there are alternatives to the sequence of the remaining mathematical code sub–units. The sequence

D - 3.10
D - 3.22
I - 3.22
D - 3.21

might be chosen by some teachers.

In the 20 recorded class sessions there were a total of 223 content cycles, each averaging 87 lines in duration (about four minutes). In 134 of these cycles, there was also no change in the other mathematatical codes (unlike the example above which had four sub–cycles), but the remaining multi–activity content cycles actually included more than 70% of all class discussion. The frequencies

TABLE XIX

% of moves initiating and closing mathematical code cycles by teachers and students in each class and for all classes combined

	Class					
	1	2	3	4	5	$\overline{X}$
1. Initiating						
T/STR	24	24	14	22	30	22.8
T/SOL	61	61	58	55	66	60.2
T/RES	0	0	0	2	0	.4
T/REA	7	10	6	2	3	5.6
						89.0
S/STR	1	2	6	5	0	2.8
S/SOL	4	1	10	12	0	5.4
S/RES	2	2	2	1	0	1.4
S/REA	1	0	4	1	1	1.4
						11.0
2. Closing						
T/STR	13	2	5	8	6	6.8
T/SOL	11	15	8	5	6	9.0
T/RES	1	2	3	11	0	3.4
T/REA	67	61	52	57	79	63.2
						82.4
S/STR	0	0	2	1	0	.6
S/SOL	1	0	3	7	1	2.4
S/RES	5	20	22	8	7	12.4
S/REA	2	0	5	3	1	2.2
						17.6

of transitions in mathematical activity and focus within these content cycles are given in Tables XX to XXII. In each table the data are grouped somewhat differently in order to reveal different patterns in the changes.

Table XX shows the frequency of transition from each mathematical activity. It indicates that 56% of all sub-cycles involved development of mathematical systems, 17% recall, 14% illustration, and 12% examination. As might be expected, the table shows that recall was far more likely to occur at the beginning of a cycle than at the end, and illustration and examination occurred more often near the end of a cycle.

	D	R	I	E
$\frac{\text{Frequency begin cycle}}{\text{Frequency end cycle}}$	1.05	2.56	.29	.33

Two other patterns appear in Table XX. First, some mathematical activities were far more likely to precede rather than follow certain other activities. For instance, recall preceded illustration and development about twice as often as it followed these activities. Second, another related pattern suggests that some mathematical activities were more likely to occur than others, given that a certain specific activity had just occurred. For instance, although development took up 56% of all sub-cycles, this activity was far less likely to occur following a recall sub-cycle (one of three) and more likely to occur following an examining cycle (two of three).

Another complementary pattern is suggested by Table XXI, where transition frequencies are tabulated for changes in the focus of activity. The codes for this aspect of mathematical activity were designed so that 1.1, 2.1, 3.1, and 4.1 indicate *definitions* of sets, operations, mappings, and relations, respectively; 2.21, 3.21, and 4.21 indicate specific *computational facts;* 1.2, 2.22, 3.22, 4.22, and 6.0 indicate *general properties;* and 5.0 indicates talk about *logic.* Grouping sub-cycles under these four major categories shows that the most common focus was general properties of sets, operations, mappings, and relations—47% of all sub–cycles. Definitions and computational facts were each considered in approximately 22% of the sub-cycles, and logic in 9%. This emphasis on general or quantified properties is somewhat surprising, but it apparently reflects the more theoretical, conceptual orientation of the experimental textbook.

The most obvious indication of the table is the fact that cycles began with definition far more often than they ended with that focus (a ratio of nearly 2 : 1), whereas discussions of logic occurred

TABLE XX

Transition frequencies for mathematical activity within content cycles*

Second Sub-cycle

First Sub-cycle	O	D	R	I	C	E	AN	
O		58	23	4	1	3	0	89
D	55	69	8	22	0	23	1	178 = 56%
R	9	14	11	15	0	4	0	53 = 17%
I	14	17	7	1	0	6	0	45 = 14%
C	1	0	0	0	1	0	0	2 = .7%
E	9	20	4	3	0	4	0	40 = 12%
AN	1	0	0	0	0	0	0	1 = .3%

*O is a dummy introduced at the beginning and end of each cycle; that is, (O, I) indicates a cycle begun with illustration and (E, O) indicates a cycle ended with examination.

TABLE XXI

Transition frequencies for mathematical activity focus within content cycles

Second Sub-cycle

First Sub-cycle	0	*Definition*	*Computational Facts*	*General Properties*	*Logic*	
0		26	21	37	5	
Definition	14	25	13	15	3	70 = 22%
Computational Facts	21	6	14	24	4	69 = 22%
General Properties	42	11	17	64	17	151 = 47%
Logic	12	3	4	10	0	29 = 9%

more frequently at the end of cycles than at the beginning (a ratio of more than 2 : 1). There was a trend for definition to precede computational facts and general properties, for specific facts to precede general properties, and for discussion of logic to follow rather than precede consideration of general properties. Thus, the composite of these trends is a pattern for content cycles: definition led to computation, then to general properties, and then occasionally to discussion of logical issues.

Table XXII, which indicates transition frequencies for mathematical activity and focus simultaneously, makes the common internal structures of content cycles even clearer. Content cycles commonly begin with recall of a general property (of an operation, mapping, or relation), or developing of a definition, specific fact, or general property. Depending on which of these occurred, the cycle would proceed in one of several directions. For instance, if the first activity was developing a definition, the next was likely to be either developing specific facts (35% of the time), developing general properties (20% of the time), or illustrating the definition (20% of the time. Then, for each of these new states there were two or three most likely changes. When activity entered the stage of discussion about logic (D.5), the chances were about 4 in 9 that the content cycle would end when this activity did.

In summary, the table indicates that the various mathematical activities and the focus did not simply occur with certain relative frequencies, but that likelihood of occurrence was definitely affected by the nature of the preceding activity.

Because of the limited number of multi–dimensional content cycles available for examination (only 89 for five classes), it is impossible to give a reliable characterization of transition patterns in individual classes. One possible future approach to this problem would be to use only the mathematical code portion of the system of analysis to study individual classes over a longer period of time.

Since each content cycle did not generally contain many mathematical activity sub–cycles (more than 80% contained three or fewer), it is also difficult to determine whether occurrence of a particular mathematical activity is strongly influenced by activities other than the one directly preceding. Answers to this and the preceding problem will go even further toward an accurate appraisal of the influence of mathematical structure on the structure of discourse about mathematics.

SUBORDINATE CONTENT AND MATHEMATICAL ACTIVITY

Some mathematical code cycles were interrupted by diversions from the main content or activity. To deal with such situations,

TABLE XXII

Transition frequencies within content cycles—mathematical activity and focus*

First Sub-cycle \ *Second Sub-cycle*	O	D.1	D.21	D.22	D.5	R.1	R.22	I.1	I.22	E.1	E.21	E.22
O		15	21	19	3	8	15	2	1	1	0	2
D.1	4	0	7	4	1	0	1	4	1	2	0	0
D.21	20	1	0	15	0	0	4	0	3	5	8	0
D.22	23	1	9	10	12	0	1	0	13	0	0	4
D.5	8	1	3	3	0	0	1	1	0	0	1	0
R.1	3	2	0	0	0	2	4	4	0	1	0	0
R.22	6	2	4	5	0	5	0	0	12	1	0	0
I.1	3	1	1	1	0	1	2	0	0	4	0	0
I.22	11	0	4	11	0	0	4	0	0	0	0	0
E1	3	1	5	1	0	0	1	2	0	1	0	0
E.21	1	0	4	0	0	0	1	0	1	0	0	0
E.22	2	0	0	2	0	0	1	0	1	0	0	0

*Only most frequent combinations listed.

Code used in the table: D = developing, R = recall, I = illustrating, and E = examining. .1 = (1.1 + 2.1 + 3.1 + 4.1), .21 = (2.21 + 3.21 + 4.21), .22 = (1.2 + 2.22 + 3.22 + 4.22 + 6.0), and .5 = (5.0).

mathematical content and activity were coded at a second, subordinate, level in addition to the level indicating major topic of the discussion.[6]

In the 20 recorded class sessions there were a total of 173 such subordinate content and activity cycles, representing 1,333 moves and 2,567 lines or approximately 14% of class time. The number of subordinate cycles varied greatly from class to class: class two diverting from the main topic 53 times in four days and class five only 20 times. However, in all classes, subordinate cycles were short, averaging between five and nine moves and less than 20 lines.

TABLE XXIII

Subordinate cycles as a portion of total activity in each class and for all classes combined

	Class					
	1	2	3	4	5	$\overline{X}$
% of all moves	10	21	15	12	9	13
% of all lines	10	20	14	10	7	12
moves/cycle	7.2	8.7	9.1	7.0	5.1	7.4
lines/cycle	13.5	16.9	19.2	10.8	9.7	14.0

In addition to initiating and closing most mathematical code cycles, teachers determined when discussion would divert from the main topic to some subordinate content or activity cycle. However, as Table XXIV shows, there was considerable variation from class to class in the extent of this domination. In two classes, students initiated no subordinate cycles, but in two others, they began 29% and 37% respectively. These differences are consistent with those in Table XIX, which give initiating and closing per cents for major mathematical code cycles.

As might be expected, the content and mathematical activity

emphases in subordinate cycles were somewhat different than in major mathematical code cycles. Subordinate cycles served most often for recall (45% of the moves and 39% of the lines), while development was far less common that it was in major activity. Illustration assumed a more important role in subordinate cycles than in major activity, but there was considerable variation from class to class in time devoted to illustration.

TABLE XXIV

% of subordinate content and activity cycles initiated and closed by teachers and students in each class and for all classes combined

	Class					
	1	2	3	4	5	$\overline{X}$
1. Initiate:						
teacher	100	91	71	63	100	85
student	0	9	29	37	0	15
2. Close:						
teacher	83	85	55	82	75	76
student	17	15	45	18	25	24

Discussion in subordinate mathematical cycles covered a wide range of mathematical topics, with no single topic clearly most common. However, the tendency of all classes to stick to the point, discussing only topics relevant to the system of integers, was maintained even when diversions occurred.

LOGICAL PROCESSES IN VERBAL ACTIVITY

Coding logical aspects of each move was a persistent source of problems. The ambiguities, mentioned in Chapter II, involving fact stating and conditional inferring, and the various forms of defining, could not be resolved by satisfactory operational category definitions. Of all the dimensions of classroom verbal activity, the logical purpose of moves was consistently least clear of all. However, ignoring refinements within major process categories (such as between fact stating and inferring, or defining and interpreting), the distribution of moves according to analytic, factual, evaluative, and justifying processes followed a consistent pattern from class to class and from session to session in a single class.

Approximately 10% of all moves were devoted to analytic process—statements about the meaning and use of language or symbolism. In individual classes, the per cent varied from a low of 6.7 to a high of 14.4. The analytic process moves were fairly evenly distributed between teachers and students.

About 50% of all moves were factual statements or questions, ranging from a low of 46.4% in class three to a high of 57.2% in class five. Again, factual moves were about evenly divided between teachers and students. However, earlier results suggest that this probably means that teachers solicited facts and students immediately responded with the ones desired.

The fact that the T/SOL–S/RES pattern was usually followed by a teacher reaction, is reflected in the distribution of evaluative logical process. About 25% of all moves involved evaluation—one for every two factual moves. Furthermore, teachers did the majority of evaluating—a ratio of more than 5 : 1. While the teacher-student ratios for factual and analytic moves were about the same in all classes, some teachers clearly allowed their students more opportunity to evaluate statements made in the classroom. In class five, the teacher–student ratio for evaluating was more than 20 : 1. But in class three, the ratio was less than 3 : 1. These two classes also showed the greatest contrast in the other teacher–directing versus student–directing measures, teacher–student ratios for structuring, soliciting, and reacting, and teacher–student ratios for initiating and closing cycles. The teacher in class three consistently permitted and succeeded in eliciting more student participation in directing classroom activity.

In all classes, an average of 9% of all moves involved justifying a definition, fact, or evaluation. The range was from a low of 7.2% in class four to a high of 11.1% in class five, with the moves about evenly divided among teachers and students in all classes.

The pattern of logical operations in classroom discourse is clear: more than 50% of all moves involved statements or questions of fact, 25% of all moves (mostly by teachers) evaluated these statements, and the remaining moves either justified previous statements or made some statement about the use of language.

TABLE XXVI

% of logical operations by teachers and students in each class and in all classes combined

Class

Logical Operation	1		2		3		4		5		$\overline{X}$	
	T	S	T	S	T	S	T	S	T	S	T	S
1. Analytic	3.8	3.1	7.5	6.9	6.1	5.5	4.4	4.6	4.3	2.4	5.2	4.5
2. Factual	26.5	24.6	26.9	23.5	23.3	23.1	23.3	27.3	31.4	25.8	26.3	24.9
3. Evaluative	23.0	4.4	21.9	3.0	17.4	6.3	20.2	4.9	22.1	0.7	20.9	3.9
4. Justifying	4.6	3.4	3.8	4.4	4.7	5.5	2.9	4.3	6.0	5.1	4.4	4.5

TABLE XXV

% of lines and moves devoted to each mathematical activity in subordinate cycles in each class and for all classes combined

Mathematical Activity	*Class* 1	2	3	4	5	$\overline{X}$
1. Developing						
moves	16	11	24	32	12	19.0
lines	17	14	25	37	19	22.4
2. Recall						
moves	35	49	30	37	75	45.2
lines	28	40	27	31	70	39.2
3. Illustration						
moves	17	7	44	5	4	15.4
lines	17	5	43	8	4	15.4
4. Comparison						
moves	8	3	0	18	9	7.6
lines	8	6	0	17	7	7.6
5. Examination						
moves	4	1	0	3	0	1.6
lines	3	1	0	3	0	1.4
6. Application (math.)						
moves	1	13	2	0	0	3.2
lines	3	17	5	0	0	5.0
7. Application (non-math)						
moves	19	16	0	5	0	8.0
lines	24	17	0	4	0	9.0

SUMMARY

The data presented in this chapter describe verbal activity in five classes participating in the SSMCIS. The core of this activity in all classes was a three-part exchange between teachers and students, beginning with a teacher solicitation of facts, followed by a student response stating a fact, and concluded by a teacher reaction evaluating this response. Occasionally, teachers began such exchanges with structuring moves or students added some reaction to that of the teacher. This pattern led to or was a result of the following specific patterns:

1. Teachers spoke more moves and more lines than all their students combined—a ratio of 3 : 2 in terms of moves and 3 : 1 in terms of lines.
2. Teachers dominated the pedagogical functions of structuring (80% of these moves), soliciting (95%), and reacting (83%), leaving responding as the major student activity (95%). The most common sequence of pedagogical moves was T/SOL —> S/RES —> T/REA.
3. The pattern of emphasis on mathematical content and activity followed the topical outline of the textbook chapter being studied in the classes. Recall of whole numbers was prominent in session I and gradually diminished in importance as integer multiplication was developed.
4. Within content cycles, the frequencies of transition in mathematical activity and focus indicated that these changes occurred in definitely patterned ways. These patterns also indicated the influence of mathematical structure on the structure of discourse.
5. More than 50% of all moves were statements or questions of fact—half teacher moves and half student moves. Teachers performed most of the 25% of all moves that involved evaluation, and equally divided with students the remaining justifying and analytic process moves.

Teacher influence in shaping the direction of classroom activity differed from class to class. But the difference was primarily one of degree rather than of kind. The roles of teachers and students in classroom discussion about mathematics have been described above. The implications of these findings and proposals for future research are discussed in Chapter IV.

4

SUMMARY AND RECOMMENDATIONS

The research described in Chapters I – III is part of a recently widespread interest in the classroom behavior of teachers and students. There are three immediate goals to this type of descriptive-analytic study: (1) to conceptualize the behavioral phenomena of teacher-student interaction in the classroom; (2) to develop observational techniques for describing behavioral patterns in individual classes; and (3) to use these techniques in determining behavioral profiles of individual classes and groups of similar classes.

Describing the classroom activity of a teacher and 30 students is hardly a routine task. In many classes, the action moves around the room rapidly and often occurs in several places simultaneously. Furthermore, each individual action usually serves several purposes in shaping the direction of classroom activity.

The present study has focused on *verbal* behavior of teachers and students in *mathematics* classes. This behavior was found to consist of a sequence of acts or *moves* which can be described according to source, duration, pedagogical purpose, mathematical content, and logical form. The different kinds of moves are put together in various ways as teachers and students play a "language game,"[1] whose objective is communication of information about mathematics. This language game model of classroom verbal interaction was suggested by Bellack's description of discourse in senior high school social studies classes. He found that the classroom language game proceeds according to certain definite, if not explicit, rules governing the roles of individual players (teacher and students), and he conjectured that changes in subject matter or grade level might lead to changes in the rules of the game—if not the kinds of moves.

If a descriptive–analytic study may be said to have a hypothesis, *the hypothesis of the present study was that the structure and content of mathematics exert a definite influence on the pattern of discourse in mathematics classes.* The language game in mathematics classes should have different rules and perhaps some different possible moves than the language game in social studies, literature, or science classes. Evidence in support of this hypothesis is in three forms:

First, examination of transcripts of the mathematics classes involved showed that, in addition to pedagogical purpose, substantive meaning, and logical form, each move in classroom discourse about mathematics serves a mathematical purpose by furthering development, examination, or application of mathematical systems.

Second, the definitions of logical process categories suitable for describing utterances in social studies classes did not capture the special usage of the term "logical" in mathematics. A modified set of logical process categories had to be developed.

Third, when the revised and expanded system of analysis was used to analyze verbal activity in five seventh-grade mathematics classes, patterns emerged which differed in several ways from those characteristic of the social studies classes, and also evidenced a structure imposed by the unique structure and methods of mathematics.

RULES OF THE LANGUAGE GAME IN MATHEMATICS CLASSES

Each of five mathematics classes participating in the SSMCIS and working on a chapter of multiplication of integers, was observed and tape-recorded for four days. The patterns of verbal communication reported in Chapter III are those that seem to characterize discussion *as it was* in those classes during the four recorded sessions. Although there is evidence that individual classes adopt styles that are reasonably consistent over time and are similar to those of other classes, no claim is made that these patterns are either typical of all mathematics classes, or that they are the patterns that should appear in "good" classes.

In all classes and from session to session in individual classes, teachers and students played clearly defined roles in the language game. In terms of sheer volume of activity, teachers made three moves for every two students made. The average teacher move was nearly twice as long as the average student move, with the result that teacher talk took up nearly three times as much class time as student talk.

Teachers also played the dominant role in shaping the direction of classroom discourse. They made 80% of all structuring moves, 93% of all soliciting moves, and 83% of all reacting moves. This left to students the duty of responding to teacher soliciting moves.

Teachers determined most changes in the substantive topic of discussion too. More than 90% of all content cycles were initiated by teacher moves.

While the trend toward teacher domination of classroom activity was clear and consistent, there were noticeable variations in the extent of teacher control from class to class. These variations in teacher style might prove a fruitful topic of future research.

The single most common form of logical process was statement or question of fact (51% of all moves—about half by teachers and half by students), followed in importance by evaluative statements (25% of all moves—primarily by teachers).

The patterns in both pedagogical and logical purpose of moves indicate that in the observed mathematics classes the language game consists of numerous episodes in which the teacher solicits facts, the student responds with a statement of fact, and the teacher reacts by evaluating the student response. Occasionally these episodes are initiated by teacher structuring moves which make statements about the meaning or use of language. Occasionally justification is given in addition to statements of fact or evaluation. These patterns of pedagogical and logical activity are much the same as those Bellack found in senior high school social studies classes, the only exception being that in social studies, discussion seems to consist of fewer but longer moves.

As was anticipated, the logical organization of mathematics cannot be casually rearranged in teaching. The content and mathematical activity of discourse in the observed classes was remarkably close to that contained in the textbook chapter being studied. This meant that recall of previously developed number systems was followed by definition of multiplication in the integers, and then by comparison of this operation to composition of dilation mappings.

The influence of mathematical structure on the structure of discourse was also evident in the internal structure of content cycles. During discussion of a single mathematical system, the various possible changes of mathematical activity did not occur simply at random, but with predictable frequencies that depended on the nature of previous activity. Content cycles tended to begin with recall, move on to development (of definitions, then specific facts, then general properties), and close with illustration or examination of parts of the system being studied.

In this sense, content of discussion in mathematics classes is much more predictable than in social studies classes; Bellack found wide divergence between classes in the selection of substantive topic emphasis.

FURTHER DEVELOPMENT OF THE SYSTEM OF ANALYSIS

The patterns of verbal activity detailed in Chapter III and summarized above, describe the language game as played in five classes studying a common mathematical topic. While the concepts and observational techniques developed to uncover these patterns should be useful in describing verbal activity in a much broader range of mathematics classes, this is a hypothesis that only further research can test. The concept of pedagogical move is a natural and helpful basic unit which should be generally applicable in analyzing mathematical class discussion. But it may well be necessary to devise new mathematical activity concepts to cover activity in other classes. The analysis of mathematical systems into sets, operations, mappings, relations, and logic might not be suitable for analysis of more intricate systems—although for most secondary school work it seems appropriate. However, a new content analysis is certainly necessary for each mathematical topic covered in classroom observation.

One aspect of the coding system that must be modified is the set of logical process categories. In the present system, analytic process corresponds to the logical activity of defining, fact stating to asserting a proposition to be true or false($\vdash p$ or $\vdash \sim p$), and inference and deductive justification together to the logical act of deduction. From a formal logical point of view, evaluation and fact stating are the same activity. They both involve statements about the truth of a proposition (except the rare occasion when evaluation refers to good or bad), but occurrence in classroom discourse indicates quite different roles for the two. For instance, in the exchange

T(Fac): What is 7 times 2?
S(Fac): 14.
T(Val): 14 is right.

the second teacher move clearly rates the student response; it is not primarily intended as an assertion of fact. The difficulty illustrated here is a basic problem in descriptive–analytic studies: namely, finding theoretical concepts which accurately convey the meaning of observed behavior. It may be that use of the term "logical" process is inappropriate and that "cognitive" process might be more accurate. Perhaps logical process must be coded only for a set of consecutive related moves and not for individual moves.

Another alternative is to relate use of the term "logical" to the study of mathematics as a deductive science. A deductive science consists of certain undefined terms, axioms, definitions, and theorems, and each move in a mathematics class can be viewed as a step in the development, examination, or application (at varying levels of formality) of a deductive system. Thus, the logical aspect of any move could be considered to be the part of a deductive system being considered.

The usefulness of this or any such theoretical framework can be determined only by trying to describe actual classroom behavior using these concepts. Then the entire system of analysis must be tested for validity and reliability in a wider range of classes.

In any case, each move in the classroom language game serves many purposes, of which the major coding dimensions—pedagogical, mathematical, and logical—are only three.

RESEARCH SUGGESTED BY THIS STUDY

The limitations of the present study with respect to grade level, number of classes observed, period of class time observed, and mathematical topic of discussion, suggest many possibilities for further investigation. For instance:

1. Are the patterns typical of these classes common in eleventh- or twelfth-grade mathematics classes?
2. Do variations in the mathematical topic under discussion produce variations in the pattern of mathematical activity or logical process?
3. Do individual curricula, as embodied in textbooks, induce distinguishable classroom teaching styles?[2]
4. Are there teaching styles, as measured by teacher-student roles in the classroom language game, which are optimal for producing student learning?
5. What interrelationships are there between the findings of such a cognitive-oriented examination of classroom communication and affect-oriented research such as that of Flanders?[3]

6. Can the behavioral concepts identified in the descriptive study of teachers be used to construct models of teaching behavior, the effectiveness of which can then be compared experimentally?[4]

The research procedures employed in this study—tape-recording, transcribing, coding, and analyzing—are extremely time-consuming and tedious. They seemed a necessary phase of the search for new behavioral variables in classroom verbal interaction, but the results suggest that future research can be simplified in several ways.

For instance, any one of the three major coding dimensions (pedagogical moves, mathematical activity, logical process) might be used by observer-coders who make instantaneous classifications of the purposes of classroom behavior. This would be particularly feasible with the mathematical codes, which change less often than others, and it would permit observation of many more content and activity cycles. As a caution, however, experience indicates that adequate analysis of the mathematical aspects of discussion, even for tape-recorded sessions, demands coders who are thoroughly familiar with the content and structure of mathematics.

There is another reasonable alternative to the techniques of this study. Discourse in individual classes develops a pattern of rate and distribution according to source and pedagogical purpose of moves that remains remarkably constant over time. Thus, appropriately chosen samples of discourse—coded according to source, duration, and pedagogical purpose—could serve as suitable estimates of the total pattern with respect to these variables.

OTHER PERSPECTIVES ON CLASSROOM VERBAL ACTIVITY

Conceptual metaphors, such as the language game model of classroom verbal activity, serve a dual purpose in empirical research. They help to organize many observations into a coherent explanation of some phenomenon, and they also suggest new questions worth asking or further observations that might lead to interesting results. Different metaphors suggest different research emphases, and the results of investigations prompted by different approaches often complement each other and give a more complete picture of the phenomenon being studied.

For example, examining classroom verbal activity from the perspective of Shannon and Weaver's communication theory model[5] would offer some insight into classroom verbal activity not considered by the language game model used in the present study. Instead of looking for the rules of the game, one seeks the rules for coding information into transmittable messages. Instead of optimal rules for teacher-student play, one tries to discover coding procedures which succeed in sending an optimal flow of information over a given channel. Furthermore, the communication theory model suggests the concepts of *noise* in a channel and the consequent need for *redundancy* in coding, both of which have counterparts in classroom discourse and are not dealt with by the language game model.

Research from these and other conceptual points of view is essential for broad understanding of classroom interaction. The classroom is a complex, busy source of activity, much of it verbal and overt, an even greater amount silent and private. The observational techniques and the description of activity in mathematics classes produced in this study are but one, hopefully fruitful, approach to a partial description and eventual understanding of this activity.

BIBLIOGRAPHY

Amidon, Edmund, and Ned A. Flanders. "The Effect of Direct and Indirect Teacher Influence on Dependent-Prone Students Learning Geometry," *Journal of Educational Psychology,* LII (December, 1961), pp. 286-291.

———, and Elizabeth Hunter. *Improving Teaching.* New York: Holt, Rinehart and Winston, Inc., 1966. 221 pp.

Barr, Arvil S. "The Nature of the Problem," *Journal of Experimental Education,* XXX (September, 1961), pp. 5-9.

Bellack, Arno A. *The Language of the Classroom, Part II.* New York: Institute of Psychological Research, Teachers College, Columbia University, 1965. 259 pp.

———(ed.). *Theory and Research in Teaching.* New York: Teachers College Press, 1963. 122 pp.

———, and Joel R. Davitz. *The Language of the Classroom, Part I.* New York: Institute of Psychological Research, Teachers College, Columbia University, 1963. 200 pp.

———, *et. al. The Language of the Classroom.* New York: Teachers College Press, 1966. 274 pp.

Biddle, Bruce J., and Raymond S. Adams. *An Analysis of Classroom Activities, Final Report.* U. S. Office of Education, Cooperative Research Project No. 3-20-002. Columbia: University of Missouri, 1967. 715 pp.

———, and William J. Ellena (eds.). *Contemporary Research on Teacher Effectiveness.* New York: Holt, Rinehart and Winston, Inc., 1964. 352 pp.

Bruner, Jerome S. "Needed: A Theory of Instruction," *Educational Leadership,* XX (May, 1963), pp. 523-532.

Carpenter, Clarence R. "A Suggested Approach to a Systematic Analysis of Teacher Learning Operations," *Psychological Problems and Research Methods in Mathematics Training,* pp. 4-10. Rosalind L. Feierabend and Philip H. DuBois (eds.). 213 pp.

Characteristics of Mathematics Teachers That Affect Students' Learning. U. S. Office of Education, Cooperative Research Project No. 1020. Minneapolis: Minnesota School Mathematics and Science Center, Institute of Technology, University of Minnesota, 1966. 115 pp.

Chomsky, Noam. *Syntactic Structures.* The Hague: Mouton and Company, 1957. 108 pp.

Copi, Irving M. *Introduction to Logic.* New York: The Macmillan Company, 1954. 472 pp.

Cyphert, Frederick R., and Ernest Spaights (eds.). *An Analysis and Projection of Research in Teacher Education* U. S. Office of Education, Cooperative Research Project No. F-105. Columbus: The Ohio State University Research Foundation, 1964. 318 pp.

Davis, Robert B. "The Madison Project's Approach to a Theory of Instruction," *Journal of Research in Science Teaching,* II (June, 1964), pp. 146-162.

Eisenson, Jon, *et. al. The Psychology of Communication.* New York: Appleton-Century-Crofts, 1963. 394 pp.

Exner, Robert M., and Myron F. Rosskopf. *Logic in Elementary Mathematics.* New York: McGraw-Hill Book Company, Inc., 1959. 274 pp.

Feierabend, Rosalind L., and Philip H. DuBois (eds.). *Psychological Problems and Research Methods in Mathematics Training.* St. Louis: Washington University, 1959. 213 pp.

Flanders, Ned A. *Teacher Influence, Pupil Attitudes, and Achievement.* Cooperative Research Monograph No. 12. Washington: U. S. Department of Health, Education and Welfare, U. S. Office of Education, 1965. 126 pp.

———, and Edmund Amidon. *The Role of the Teacher in the Classroom: A Manual for Understanding and Improving Teachers' Classroom Behavior.* Minneapolis: Paul S. Amidon and Associates, 1963. 68 pp.

Gage, Nathaniel L. (ed.). *Handbook of Research on Teaching.* Chicago: Rand McNally and Company, 1963. 1218 pp.

———. "Research on Cognitive Aspects of Teaching." *The Way Teaching Is,* pp. 29-44. Mrs. Curtice Hitchcock (ed.). A Report Prepared by the Seminar on Teaching. 80 pp.

———. "Toward a Cognitive Theory of Teaching," *Teachers College Record,* LXV (February, 1964), pp. 408-412.

Getzels, Jacob W., and Philip Jackson. "Research on the Variable Teacher: Some Comments," *School Review,* LXVIII (Winter, 1960), pp. 450-462.

Henderson, Kenneth B. "A Model for Teaching Mathematical Concepts," *The Mathematics Teacher,* LX (October, 1967), pp. 573-575.

Hitchcock, Mrs. Curtis (ed.). Association for Supervision and Curriculum Development. *The Way Teaching Is.* A Report Prepared by the Seminar on Teaching. Washington: Association for Supervision and Curriculum Development and the Center for the Study of Instruction of the National Educational Association, 1966. 80 pp.

Hughes, Marie M., *et. al. Assessment of the Quality of Teaching in Elementary Schools: A Research Report.* U. S. Office of Education, Cooperative

Research Project No. 353. Salt Lake City: University of Utah, 1959. 314 pp.

Interaction Analysis to Study Pupil Involvement and Mathematical Content in Classrooms of the Five State Project, Training Booklet. Minnesota National Laboratory, Minnesota State Department of Education, Code No. XXXVII-B-242. St. Paul: Minnesota National Laboratory, Minnesota State Department of Education, 1964. 41 pp.

Jackson, Philip W. "The Way Teaching Is." *The Way Teaching Is,* pp. 7-27. Mrs. Curtice Hitchcock (ed.). A Report Prepared by the Seminar on Teaching. 80 pp.

Kliebard, Herbert M. "The Observation of Classroom Behavior—Some Recent Research." *The Way Teaching Is,* pp. 45-76. Mrs. Curtice Hitchcock (ed.). A Report Prepared by the Seminar on Teaching. 80 pp.

Krumboltz, John D. (ed.). *Learning and the Educational Process.* Chicago: Rand McNally and Company, 1965. 277 pp.

Macdonald, James B. "The Nature of Instruction: Needed Theory and Research," *Educational Leadership,* XXI (October, 1963), pp. 5-7.

Medley, Donald M., and Harold E. Mitzel. *Studies of Teacher Behavior: Refinement of Two Techniques for Observing Teachers' Classroom Behaviors.* Research Publication No. 28. New York: Division of Teacher Education, Board of Higher Education of the City of New York, 1955. 42 pp.

———. "The Scientific Study of Teacher Behavior," *Theory and Research in Teaching,* pp. 79-90. Arno A. Bellack (ed.). 122 pp.

———. "Measuring Classroom Behavior by Systematic Observation," *Handbook of Research on Teaching,* pp. 247-328. Nathaniel L. Gage (ed.). 1218 pp.

Nagel, Ernest. *The Structure of Science.* New York: Harcourt, Brace and World, Inc., 1961. 618 pp.

Osgood, Charles E., *et. al. The Measurement of Meaning.* Urbana: University of Illinois Press, 1957. 342 pp.

Ryans, David G. "Teacher Behavior Theory and Research: Implications for Teacher Education," *Journal of Teacher Education,* XIV (September, 1963), pp. 274-293.

Sanders, Norris M. *Classroom Questions: What Kinds?* New York: Harper and Row, 1966. 176 pp.

Scandura, Joseph M. "Teaching—Technology or Theory," *American Educational Research Journal,* III (March, 1966), pp. 139-146.

Secondary School Mathematics Curriculum Improvement Study, *Mathematics I,* Experimental Edition. New York: Teachers College Press, 1966. 264 pp.

Shannon, Claude E., and Warren Weaver. *The Mathematical Theory of Communication.* Urbana: University of Illinois Press, 1964. 117 pp.

Smith, Bunnie O. "A Concept of Teaching," *Language and Concepts in Education,* pp. 86-101. Bunnie O. Smith and Robert H. Ennis (eds.). 221 pp.

———. "A Conceptual Analysis of Instructional Behavior," *Journal of Teacher Education,* XIV (September, 1963), 294-298.

———, and Robert Ennis (eds.). *Language and Concepts in Education.* New York: Rand McNally and Company, 1961. 221 pp.

———, and Milton O. Meux. *A Study of the Logic of Teaching.* Urbana: Bureau of Educational Research, College of Education, University of Illinois, 1963. 231 pp.

———, *et. al. A Tentative Report on Strategies of Teaching.* U. S. Office of Education, Cooperative Research Project No. 1640. Urbana: College of Education, University of Illinois, Bureau of Educational Research, 1964.

Solomon, Daniel, *et. al. "Dimensions of Teacher Behavior," Journal of Experimental Education,* XXXIII (Fall, 1964), pp. 23-40.

Solomon, Herbert (ed.). *Mathematical Thinking in the Measurement of Behavior.* Glencoe: The Free Press, 1960. 314 pp.

Stolurow, Lawrence M. "Model the Master Teacher or Master the Teaching Model," *Learning and the Educational Process,* pp. 223-247. John D. Krumboltz (ed.). 277 pp.

Taylor, James G. *The Behavioral Basis of Perception.* New Haven: Yale University Press, 1962. 379 pp.

Toda, Masanao. "Information-Receiving Behavior of Man," *Psychological Review,* LXIII (May, 1956), pp. 204-212.

Travers, Robert M. W. "Taking the Fun Out of Building a Theory of Instruction," *Teachers College Record,* LXVIII (October, 1966), pp. 49-60.

———. "The Utilization in Education of Knowledge Derived From Research on Learning," *An Analysis and Projection of Research in Teacher Education,* pp. 219-252. Frederick Cyphert and Ernest Spaights (eds.). 318 pp.

Wingo, G. Max. "Methods of Teaching," *Encyclopedia of Educational Research* (3rd ed.), pp. 848-861. Chester W. Harris (ed.). 1564 pp.

Wright, Elisabeth Muriel J. "Development of an Instrument for Studying Verbal Behaviors in a Secondary School Mathematics Classroom," *Journal of Experimental Education,* XXVIII (December, 1958), pp. 103-121.

———. "A Rationale for Direct Observation of Verbal Behaviors in the Mathematics Classroom," *Psychological Problems and Research Methods in Mathematics Training,* pp. 38-46. Rosalind L. Feierabend and Philip H. DuBois (eds.). 213 pp.

———. *Teacher-Pupil Interaction in the Mathematics Classroom.* A Subproject Report of the Secondary Mathematics Evaluation Project. Minnesota National Laboratory, Minnesota State Department of Education, Technical Report No. 76-5. St. Paul: Minnesota National Laboratory, Minnesota State Department of Education, 1967. 36 pp.

———, and Virginia H. Proctor. *Systematic Observation of Verbal Interaction as a Method of Comparing Mathematics Lessons.* U. S. Office of Education, Cooperative Research Project No. 816. St. Louis: Washington University, 1961. 224 pp.

APPENDIX

INSTRUCTIONS FOR CODING

The system for analyzing protocols is described in Chapter II. However, the reader who wishes to use the system is advised to familiarize himself with several related studies in order to appreciate the objectives and methods of protocol analysis in descriptive-analytic studies of teaching. The most pertinent of these is:

Bellack, Arno A. *et. al. The Language of the Classroom.* (especially chapters 1, 2, 4-6)

Also suggested are:

Smith, Bunnie O. *et. al. A Study of the Logic of Teaching.* (especially chapter 1)

Wright, Elisabeth Muriel J., and Virginia Proctor. *Systematic Observation.*

The following instructions are specific coding procedures found helpful in applying the system of analysis to actual protocols.

GENERAL REMARKS

1. Coding is done from the viewpoint of a non-participating observer who infers the meaning of an utterance from the speaker's verbal behavior.
2. Grammatical form may give a clue to meaning, but it is not

decisive in coding. For example, SOL may be found in declarative, interrogative, or imperative utterances, such as:

T/SOL: Give me the answer to 3(b). . .

and RES may be in the form of a question

S/RES: Is it 23?

when the student response is tentative.

3. Each utterance is coded in context since meaning can be judged accurately only by considering the effect of the move as part of continuous discourse. In particular, the content, mathematical activity, and logical processes of SOL moves are coded according to the response expected by the solicitation. For example:

   T/SOL: Do you remember what 7 times 13 is?

   is coded Wm/R2.21/Fac because it calls for recall of a fact from Wm.
4. Individuals (usually T) commonly make two or three moves in a single uninterrupted utterance. However, partitioning utterances in order to code every nuance of meaning is extremely difficult. Each structuring, soliciting, responding, or reacting move should be coded by judging dominant communicative effect.
5. If any dimensions of a move are uncodable (because of undecipherable recordings or confusion on the part of the speaker himself), those dimensions are coded NOC. Generally, mathematical content and activity are assumed to remain constant over an uncodable move.

CODING MOVES

1. Structuring moves are those that turn the discussion in a new direction without soliciting any specific response. However, STR moves are often directly followed by a related SOL. In fact, the last sentence of a STR can often be interpreted as part of the succeeding SOL. In such cases the following rules of coding are used:
   (a) When a SOL is preceded by a long (two or more complete sentences) STR, the SOL begins with the first soliciting sentence.
   (b) When a SOL is preceded by a single sentence of structur-

ing character, the preceding sentence is coded STR if the SOL is independent of its content. If the SOL has meaning only in connection with the preceding sentence, SOL is coded for the combined unit. For example:

T/SOL: Suppose x is 2 and y is 3. What is x - y?
T: Let's turn to addition of integers. (STR)/What is -7 + 3? (SOL)
T: If we look at integers, the story is different. Suppose a is 3 and b is 1. (STR)/What is the solution of a + x = b? (SOL)

2. Assignments are coded SOL.
3. When a speaker answers his own question immediately, the question is considered rhetorical, not SOL. However, if the answer comes only after a pause, it is assumed that the answer is a reaction to the lack of a response. If the teacher makes a SOL and, receiving no answer, re-addresses the SOL, a single SOL is coded. Similarly, if the teacher asks two distinct questions at once, a single SOL is coded.
4. Often a student utterance occurs in response to a tacit question (as in giving answers to homework problems) or a question written on the chalkboard. Such moves are coded RES-N (a code not accounted for in the data analysis). Often a speaker (usually T) asks a question, receives a response, evaluates the response and addresses the same question to another person by simply naming the next responder. The second SOL is coded the same as the first (except perhaps for duration). For example:

   T: What is the distributive law, John? (SOL)
   S: a(b + c) = ab + c. (RES)
   T: No, that's not it. (REA) / Mary? (SOL)

   However, if the teacher merely gives a student the right to speak, making no tacit SOL, the teacher's utterance is disregarded. For example:

   S: I think it is 5. (RES)
   T: Jim? (omit from coding)
   S: Well if it is 5 then the other answer is wrong. (REA)

5. Occasionally a move is interrupted by another move (or moves) and then resumed. In such a case the resumption is coded as the original with the postscript -M to indicate a mediated move. For example:

   S: Well if you multiply both sides by 7. . . (RES)

T: All right. (REA)
S: . . . You get x = 28. (RES-M)

6. Responses usually occur directly after the solicitation they answer, but when the pertinent SOL appears earlier in the discourse, the RES is coded with the postscript M and a number to indicate the location of the SOL. For example:

   T: What is 7 + -3? (SOL)
   S: 10. (RES)
   S: 4. (RES-M2)

7. Reacting moves, although usually clear in intent at the outset, often seem to lead discussion beyond the specific remarks which provoked the reaction. A continuous REA is coded when the speaker discusses (not necessarily the exactly preceding move), rephrases or interprets or expands a previous move by drawing implications from it. An REA ends when the utterance ends and there is a distinct shift in content or discussion by some verbal structuring form such as "Now let's look at . . ."
8. When a reaction to a move (often a Val) is followed by remarks summarizing a segment of the preceding discussion, two REA moves are coded even though occuring together in a single utterance. However, the second REA must clearly represent a shift in purpose (usually logical process) and serve as a summary, not merely an expansion of the preceding discussion.
9. Reaction to a solicitation occurs only when the speaker remarks on the pertinence or form of a SOL. For example:

   S: Put parentheses in there. (SOL)
   T: Good. I'm glad to see you are punctuation conscious. (REA)

CODING CONTENT AND MATHEMATICAL ACTIVITY

1. Because content and mathematical activity are coded at two levels of specificity, it is often necessary to examine several pages of a protocol to determine appropriate general and specific codes. In deciding whether an apparent shift in content or activity represents entry into a new cycle, or simply diversion to a contributing sub-cycle, the following guidelines are helpful:
   (a) Does the discussion return to the original content and mathematical activity within a reasonable length of time? When the return occurs, is there some verbal indication that the discussion is now picking up where it left off? If

so, it is coded as a sub-cycle. For example:

T: We would like to have a cancellation property in Z./ Does someone recall the cancellation law in W?
S: ab = ac implies b = c.
T: Okay./ And we want multiplication to be commutative in Z.

This segment is coded as a three-move sub-cycle within a developing cycle.

(b) Does the proposed sub-cycle constitute a coherent unit independent of, yet contributing to, the main discussion? If not, it is coded as part of the main discussion, not as a sub-cycle. Conversely, can the main discussion be understood clearly if the proposed sub-cycle is removed altogether? If not, the entire segment is coded at one level. A common source of such coding difficulty is developing activity where recall may occur as part of a proof. For example:

T: To prove that 5(-5) = -25, we look at 5 + (-5). What is 5 + (-5)?
S: O.
T: It is 0. Now we multiply both sides by 5, and what do we get?
S: 25 + 5(-5) = 0.
T: So 5(-5) must be?
S: -25,

is coded as a single level cycle Zm/D2.21, even though recall of an addition fact is part of the proof. The recall segment is an integral part of the proof.

2. When a move is coded AM or AN, the system being applied as content is indicated, but no specific activity code, such as 2.1, 3.22, is.
3. In determining whether a discussion involves development of a theorem or simply a general property of operations, mappings, or relations, these rules should be followed:
   (a) "=" is considered a part of logic, not a defined relation like order or perpendicularity. Therefore, "a + b = b + a" is coded 2.22 and not 6.0.
   (b) A statement (such as the distributive property) involving one or more operations is still considered 2.22.
   (c) In general, the code which most accurately represents the meaning of the statement being established is chosen.
4. In comparing, C is coded if it is a general discussion of sys-

tems. Comparing is preferred to recall when both are applicable.

CODING LOGICAL PROCESS

1. Since logical process commonly varies from move to move, coding according to the context is often more difficult than with mathematical activity. Thus, before attempting to make fine distinctions (such as between the varieties of justifying or the two kinds of factual process), it is often helpful to proceed by elimination in order to determine the major process category.
2. When a move includes two or more logical processes, coding is by which is the dominant logical function or, if this is not clear, according to the following order of preference:
   a. Justifying
   b. Evaluating (when positive or negative rating)
   c. Factual
   d. Analytic
   e. Evaluating (when simply acknowledging without positive or negative rating)
3. Analytic (as opposed to factual process) is coded only when the discussion clearly centers around usage of terminology or symbolism, or when the statement is clearly intended as clarification by rephrasing. For example:

   T: Who can state the associative property for addition using a, b, c?

   is primarily a solicitation of fact, not interpretation.
4. Two logical process distinctions are particularly difficult to make. The first is between evaluative and fact stating moves. The second is between the factual processes Fac and Inf.
   (a) Grammatical form of a SOL often suggests that a speaker wants someone to determine whether a given statement is true or false (e.g., T/SOL: Do you think addition is an operation in Z?). However, the response can also be viewed as fact stating (S/RES: Yes, addition is an operation in Z.). In general, Val seems most appropriate coding for moves reacting to a statement of fact, not for moves responding to a "Yes" or "No" type question. Therefore, unless moves clearly rate a previously stated fact, they are coded Fac rather than Val when the choice is between these two processes.
   (b) Conditional inference is the process of completing an "if

p then q" statement, given the hypothesis p. From certain given conditions, a speaker provides or asks for some logical consequence. There are many situations in which a SOL has the "If p then?" form, but the expected response is really only statement of a fact. For example:

T/SOL: If r = 4 and s = -3, then rs is?
S/RES: -12,

suggests that Inf be coded only in situations where (1) several possible correct answers can be given, or (2) the expected conclusion is a newly discovered fact, arrived at by reasoning. For example:

T/SOL: If we know that positive times negative is negative, what should negative times negative be? (Inf)
S/RES: Negative./ (Inf, even though false)

5. In the Bellack study, *explaining* was viewed as the same logical process as *conditional inference*—only viewed from the opposite direction. Inference gives the conclusion that follows from certain conditions; explanation gives the conditions that will lead to a certain conclusion. For example:

   S/SOL: Tell me how you got -5 for that one.

   and

   T/STR: To do these, you square the first number and then add the second.

   involve explanations. However, the two explanations will really serve different purposes. The first SOL asks primarily for a justification; the second STR gives directions on how to arrive at a given answer. Moves that solicit or give explanations occur as analytic, factual, and justifying processes. Thus, they should (and in most cases can) be coded as Aly, Fap, or Jus.
6. Key words, such as "why," "what," or "how," often give a clue to logical process coding.
7. When a SOL calls for justification, code Jus rather than a more specific code unless context indicates the particular kind of justification expected.

FOOTNOTES

CHAPTER I

[1]Philip W. Jackson, "The Way Teaching Is," *The Way Teaching Is,* Report of the Seminar on Teaching, Association for Supervision and Curriculum Development and the Center for the Study of Instruction (Washington, D.C.: Association for Supervision and Curriculum Development and the National Education Association, 1966), pp. 7-27.

[2]Robert M. Travers, "Taking the Fun out of Building a Theory of Instruction," *Teachers College Record,* LXVIII (October 1966), pp. 49-60.

[3]B.O. Smith *et. al., A Study of the Logic of Teaching* (Urbana: Bureau of Educational Research, College of Education, University of Illinois, 1963), pp. 1-8.

[4]Robert B. Davis, "The Madison Project's Approach to a Theory of Instruction," *Journal of Research in Science Teaching,* II (June 1964), pp. 146-162.

[5]Travers, "The Utilization in Education of Knowledge Derived from Research on Learning," *An Analysis and Projection of Research in Teacher Education* (Columbus, Ohio: The Ohio State University Research Foundation, 1964), p. 235.

[6]Jerome S. Bruner, "Needed: A Theory of Instruction," *Educational Leadership,* XX (May 1963), p. 524.

[7]G. Max Wingo, "Methods of Teaching," *Encyclopedia of Educational Research* (Third Edition, New York: Macmillan and Company, 1960), p. 850.

[8]*Ibid.*

[9]C.R. Carpenter, "A Suggested Approach to a Systematic Analysis of Teacher-Learning Operations," *Psychological Problems and Research Meth-*

ods in Mathematics Training, U.S. Department of Health, Education and Welfare, Office of Education, Cooperative Research Project No. 642 (St. Louis: Washington University, 1959), p. 4.

[10]Jackson, *op. cit.,* p. 9.

[11]Bruce Biddle, "The Integration of Teacher Effectiveness Research," *Contemporary Research on Teacher Effectiveness* (New York: Holt, Rinehart and Winston, Inc., 1964), pp. 9-10.

[12]*Characteristics of Mathematics Teachers that Affect Students' Learning,* U.S. Department of Health, Education and Welfare, Cooperative Research Project No. 1020 (Minneapolis: Minnesota School Mathematics and Science Center, Institute of Technology, University of Minnesota, 1966), pp. 104-109.

[13]*Ibid.,* p. 109.

[14]Smith, *op. cit.,* p. 2.

[15]Elisabeth Muriel Wright, "Development of an Instrument for Studying Verbal Behaviors in a Secondary School Mathematics Classroom," *Journal of Experimental Education,* XXVIII (December 1959), p. 103.

[16]Investigations of classroom teacher behavior have been traced as far back as 1914. For a review of this early research see Donald M. Medley and Harold E. Mitzel, "Measuring Classroom Behavior by Systematic Observation," *Handbook of Research on Teaching* (Chicago: Rand McNally and Co., 1964), pp. 254-297. However, these same authors conclude elsewhere that most such investigations prior to about 1950 are lightly regarded because criteria of effectiveness used are suspect (ratings of supervising teachers being a chief measure of effectiveness), and because observational data were not adequately objective. This argument is detailed in Medley and Mitzel's "The Scientific Study of Teacher Behavior," *Theory and Research in Teaching* (New York: Teachers College Press, 1963).

[17]David G. Ryans, "Teacher Behavior Theory and Research: Implications for Teacher Education," *Journal of Teacher Education,* XIV (September 1963), p. 292.

[18]Nathaniel L. Gage, "Research on Cognitive Aspects of Teaching," *The Way Teaching Is,* p. 31.

[19]For a detailed discussion of the background of this type of research see Ned A. Flanders, "Teacher Influence in the Classroom," *Theory and Research in Teaching,* pp. 37-52.

[20]Ned A. Flanders, *Teacher Influence, Pupil Attitudes, and Achievement,* U.S. Department of Health, Education and Welfare, Office of Education, Cooperative Research Monograph No. 12 (Washington, D.C., 1965), p. 7.

[21]*Ibid.,* p. 108.

[22]See Edmund Amidon and Elizabeth Hunter, *Improving Teaching* (New York: Holt, Rinehart and Winston, Inc., 1966); or Ned A. Flanders and Edmund Amidon, *The Role of the Teacher in the Classroom—A Manual for Understanding and Improving Teacher's Classroom Behavior* (Minneapolis: Paul S. Amidon and Associates, 1963).

[23]Gage, *op. cit.,* p. 32.

[24]Others are discussed in Herbert M. Kliebard, "The Observation of Classroom Behavior—Some Recent Research," *The Way Teaching Is,* pp. 45-76.

[25]Smith, *op. cit.,* p. 14.

[26]Smith, *et al., A Tentative Report on the Strategies of Teaching,* U.S. Department of Health, Education and Welfare, Office of Education, Cooperative Research Project No. 1640 (Urbana: Bureau of Educational Research, College of Education, University of Illinois, 1964).

[27]Arno A. Bellack *et al., The Language of the Classroom* (New York: Teachers College Press, 1966).

[28]Wright, *op. cit.,* p. 103.

[29]Wright and Virginia H. Proctor, *Systematic Observation of Verbal Interaction as a Method of Comparing Mathematics Lessons,* U.S. Office of Education, Cooperative Research Project No. 816 (St. Louis: Washington University, 1961).

[30]Kenneth B. Henderson, "A Model for Teaching Mathematical Concepts," *The Mathematics Teacher,* LX (October 1967), pp. 573-575.

[31]Wright, *Teacher-Pupil Interaction in the Mathematics Classroom,* A Subproject Report of the Secondary Mathematics Evaluation Project, Technical Report No. 67-5, Minnesota National Laboratory, Minnesota State Department of Education (St. Paul, 1967), p. I-2.

[32]The move definitions are adapted from Bellack, *op. cit.,* pp. 16-19.

CHAPTER II

[1]Claude Shannon and Warren Weaver, *The Mathematical Theory of Communication* (Urbana: University of Illinois Press, 1964), p. 8.

[2]Bellack, *op. cit.,* pp. 1-4.

[3]*Ibid.,* pp. 16-19.

[4]*Ibid.,* pp. 16-17.

[5]*Ibid.,* p. 18.

[6]*Ibid.*

[7]*Ibid.,* pp. 18-19.

[8]Secondary School Mathematics Curriculum Improvement Study, *Mathematics I* (Experimental Edition) (New York: Teachers College Press, 1966), pp. 105-115.

[9]Robert M. Exner and Myron F. Rosskopf, *Logic in Elementary Mathematics* (New York : McGraw-Hill Book, Co., Inc., 1959), p. 102.

[10]Smith, *op. cit.,* pp. 3-4.

[11]Smith, *op. cit.,* pp. 66-200; Bellack, *op. cit.,* pp. 22-26.

[12]Bellack, *op. cit.,* p. 22.

[13]*Ibid.,* p. 23.

[14]*Ibid.,* p. 24.

[15]*Ibid.*

[16]This method of defining cycles in classroom discourse was suggested by Biddle and Raymond S. Adams, *An Analysis of Classroom Activities,* a

Final Report, U.S. Department of Health, Education and Welfare, Office of Education, Contract No. 3-20-002 (Columbia: University of Missouri, 1967).

[17]Bellack, *op. cit.*, p. 36.

CHAPTER III

[1]Bellack, *op. cit.*, pp. 46-48.

[2]*Ibid.*

[3]*Ibid.*, pp. 56-67.

[4]*Ibid.*, pp. 68-69.

[5]See chapter II for a detailed explanation of the way cycles are identified.

[6]See chapter II for a detailed explanation of this coding.

CHAPTER IV

[1]Bellack, *op. cit.*, p. 237.

[2]This is the type of question investigated by the Minnesota Study mentioned in chapter I: Wright, *Teacher-Pupil Interaction.*

[3]Flanders, *op. cit.*

[4]This is the type of question explored by Henderson, *op. cit.*, pp. 573-575. It is also the approach to teaching favored by Lawrence M. Stolurow, "Model the Master Teacher or Master the Teaching Model," *Learning and the Educational Process* (Chicago: Rand McNally and Co., 1965), pp. 223-247.

[5]Shannon and Weaver, *op. cit.*, p. 8.

DATE DUE